FALSE ALARM

FALSE ALARM

HOW CLIMATE CHANGE PANIC COSTS US TRILLIONS, HURTS THE POOR, AND FAILS TO FIX THE PLANET

BJORN LOMBORG

BASIC BOOKS

New York

Basic Books
Hachette Book Group
1290 Avenue of the Americas, New York, NY 10104
www.basicbooks.com

Printed in the United States of America

First Edition: July 2020

Published by Basic Books, an imprint of Perseus Books, LLC, a subsidiary of Hachette Book Group, Inc. The Basic Books name and logo is a trademark of the Hachette Book Group.

The Hachette Speakers Bureau provides a wide range of authors for speaking events. To find out more, go to www.hachettespeakersbureau.com or call (866) 376-6591.

The publisher is not responsible for websites (or their content) that are not owned by the publisher.

Editorial production by Christine Marra, *Marra*thon Production Services. www.marrathoneditorial.org
Interior book design by Jane Raese
Set in 10-point Utopia

Library of Congress Cataloging-in-Publication Data has been applied for.

ISBN 978-1-5416-4746-6 (hardcover), ISBN 978-1-5416-4748-0 (ebook)

LSC-C

10   9   8   7   6   5   4   3   2   1

CONTENTS

SECTION 4
HOW TO FIX CLIMATE CHANGE

SECTION 5
TACKLING CLIMATE CHANGE AND
ALL THE WORLD'S OTHER CHALLENGES

FALSE ALARM

SECTION ONE

CLIMATE OF FEAR

INTRODUCTION

WE LIVE IN AN AGE OF FEAR—particularly a fear of climate change. One picture summarizes this age for me. It is of a girl holding a sign saying:

You'll die of old age
I'll die of climate change

This is the message that the media is drilling into our heads: climate change is destroying our planet and threatens to kill us all. The language is of apocalypse. News outlets refer to the "planet's imminent incineration" and analysts suggest that global warming could make humanity extinct in a few decades. Recently, the media has informed us that humanity has just a decade left to rescue the planet, making 2030 the deadline to save civilization. And therefore we must radically transform every major economy to end fossil fuel use, reduce carbon emissions to zero, and establish a totally renewable basis for all economic activity.[1]

Children live in fear and line the streets in protest. Activists are cordoning off cities and airports to raise awareness that the entire population of the planet is facing "slaughter, death, and starvation."[2]

Influential books reinforce this understanding. In 2017, journalist David Wallace-Wells wrote a lengthy and terrifying description of global warming impacts for *New York* magazine. Although the article was generally panned by scientists as exaggerated and misleading, he went on to publish the same argument in book form in *The Uninhabitable Earth*, which became a bestseller. The book revels in unabashed alarmism: "It is worse, much worse, than you think." Likewise, in his 2019 book, *Falter*, naturalist Bill McKibben warned that global warming is the greatest threat to human civilization, worse even than nuclear war. It could finish off humanity not with an explosion but "with the burble of a rising ocean." A bookshelf would groan under the weight of recent books with deliberately terrifying titles and messages: *Field Notes from a Catastrophe: Man, Nature, and Climate Change; Storms of My Grandchildren: The Truth About the Coming Climate Catastrophe and Our Last Chance to Save Humanity; The Great Derangement: Climate Change and the Unthinkable;* and *This Is the Way the World Ends: How Droughts and Die-offs, Heat Waves and Hurricanes Are Converging on America.*[3]

Media outlets reinforce the extreme language by giving ample space to environmental campaigners, and by engaging in their own activism. The *New York Times* warns that "across the globe climate change is happening faster than scientists predicted." The cover of *Time* magazine tells us: "Be worried. Be very worried." The British newspaper the *Guardian* has gone further, updating its style guidelines so reporters must now use the terms "climate emergency," "climate crisis," or "climate breakdown." Global warming should be "global heating." The newspaper's editor believes "climate change" just isn't scary enough, arguing that it "sounds rather passive and gentle when what scientists are talking about is a catastrophe for humanity."[4]

Unsurprisingly, the result is that most of us are very worried. A 2016 poll found that across countries as diverse as the United Arab Emirates and Denmark, a majority of people believe that the world is getting worse, not better. In the United Kingdom and the United States, two of the most prosperous countries on the planet, an astonishing 65 percent of people are pessimistic about the future. A 2019 poll found that almost half of the world's population believes climate change likely will end the human race. In the United States, four of ten people believe global warming will lead to mankind's extinction.[5]

There are real consequences to this fear. People are deciding, for instance, not to bring children into the world. One woman told a journalist: "I know that humans are hard-wired to procreate, but my instinct now is to shield my children from the horrors of the future by not bringing them to the world." The media reinforce this choice; the *Nation* wants to know:

"How Do You Decide to Have a Baby When Climate Change Is Remaking Life on Earth?"[6]

If adults are worried silly, children are terrified. A 2019 *Washington Post* survey showed that of American children ages thirteen to seventeen, 57 percent feel afraid about climate change, 52 percent feel angry, and 42 percent feel guilty. A 2012 academic study of children ages ten to twelve from three schools in Denver found that 82 percent expressed fear, sadness, and anger when discussing their feelings about the environment, and a majority of the children shared apocalyptic views about the future of the planet. It is telling that for 70 percent of the children, television, news, and movies were central to forming their terrified views. Ten-year-old Miguel says about the future:

> There won't be as many countries anymore because of global warming, because I hear on like the Discovery Channel and science channels like in three years the world might flood from the heat getting too much.

These findings, if valid nationwide, suggest that more than ten million American children are terrified of climate change.[7]

As a result of this fear, around the world children are skipping school to protest against global warming. Why attend classes when the world will end soon? Recently, a Danish first-grader asked her teacher earnestly: "What will we do when the world ends? Where will we go? The rooftops?" Parents can find a glut of online instructions and guides with titles like *Parenting in a World Hurtling Toward Catastrophe* and *On Having Kids at the End of the World*. And so, representing her generation's genuinely held terror, a young girl holds up a sign that says "I'll die of climate change."[8]

I HAVE BEEN part of the global discussion on climate change policy for two decades, since writing *The Skeptical Environmentalist*. Throughout all this time, I have argued that climate change is a real problem. Contrary to what you hear, the basic climate findings have remained remarkably consistent over the last twenty years. Scientists agree that global warming is mostly caused by humans, and there has been little change in the impacts they project for temperature and sea level rise.[9]

The political reaction to the reality of climate change has always been flawed—this, too, I have been pointing out for decades. There are, I have argued and continue to argue, smarter ways than our present-day approach to tackle global warming. But the conversation around me has changed dramatically in recent years. The rhetoric on climate change has become ever more extreme and less moored to the actual science. Over the past twenty years, climate scientists have painstakingly increased knowledge about climate change, and we have more—and more reliable—data than ever before. But at the same time, the rhetoric that comes from commentators and the media has become increasingly irrational.

The science shows us that fears of a climate apocalypse are unfounded. Global warming is real, but it is not the end of the world. It is a manageable problem. Yet, we now live in a world where almost half the population believes climate change will extinguish humanity. This has profoundly altered the political reality. It makes us double down on poor climate policies. It makes us increasingly ignore all other challenges, from pandemics and food shortages to political strife and conflicts, or subsume them under the banner of climate change.

This singular obsession with climate change means that we are now going from wasting billions of dollars on ineffective policies to wasting trillions. At the same time, we're ignoring ever more of the world's more urgent and much more tractable challenges. And we're scaring kids and adults witless, which is not just factually wrong but morally reprehensible.

If we don't say stop, the current, false climate alarm, despite its good intentions, is likely to leave the world much worse off than it could be. That is why I'm writing this book now. We need to dial back on the panic, look at the science, face the economics, and address the issue rationally. How do we fix climate change, and how do we prioritize it amid the many other problems afflicting the world?

———

CLIMATE CHANGE IS REAL, it is caused predominately by carbon emissions from humans burning fossil fuels, and we should tackle it intelligently. But to do that, we need to stop exaggerating, stop arguing that it is now or never, and stop thinking climate is the only thing that matters. Many climate campaigners go further than the science supports. They implicitly or even explicitly suggest that exaggeration is acceptable because the cause is so important. After a 2019 UN climate science report led to over-the-top claims by activists, one of the scientist authors warned against exaggeration. He wrote: "We risk turning off the public with extremist talk that is not carefully supported by the science." He is right. But the impact of exaggerated climate claims goes far deeper.[10]

We are being told that we must do everything right away. Conventional wisdom, repeated ad nauseam in the media, is that we have only until 2030 to solve the problem of climate change. *This is what science tells us!*[11]

But this is not what science tells us. It's what politics tells us. This deadline came from politicians asking scientists a very specific and hypothetical question: basically, what will it take to keep climate change below an almost impossible target? Not surprisingly, the scientists responded that doing so would be almost impossible, and getting anywhere close would require enormous changes to all parts of society by 2030.

Imagine a similar discussion on traffic deaths. In the United States, forty thousand people die each year in car crashes. If politicians asked scientists how to limit the number of deaths to an almost impossible target of zero, one good answer would be to set the national speed limit to three miles per hour. Nobody would die. But *science* is not telling us that we must have a speed limit of three miles per hour—it only informs us that *if* we want zero dead, one simple way to achieve that is through a nationwide, heavily enforced three-mile-per-hour speed limit. Yet, it is a political decision for all of us to make the trade-offs between low speed limits and a connected society.[12]

Today, such is our single-minded focus on climate change that many global, regional, and even personal challenges are almost entirely subsumed by climate change. Your house is at risk of flooding—climate change! Your community is at risk of being devastated by a hurricane—climate change! People are starving in the developing world—climate change! With almost all problems identified as caused by climate, the apparent solution is to drastically reduce carbon dioxide emissions in order to ameliorate climate change. But is this really the best way to help?

If you want to help people in the Mississippi floodplains lower their risk of flooding, there are other policies that will help more, faster, cheaper, and more effectively than reducing carbon dioxide emissions. These could include better water management, building taller dikes, and stronger regulations that allow some floodplains to flood so as to avoid or alleviate flooding elsewhere. If you want to help people in the developing world reduce starvation, it is almost tragicomic to focus on cutting carbon dioxide, when access to better crop varieties, more fertilizer, market access, and general opportunities to get out of poverty would help them so much more, faster, and at lower cost. If we insist on invoking climate at every turn, we will often end up helping the world in one of the least effective ways possible.

WE ARE NOT on the brink of imminent extinction. In fact, quite the opposite. The rhetoric of impending doom belies an absolutely essential point: in almost every way we can measure, life on earth is better now than it was at any time in history.

Since 1900, we have more than doubled our life expectancy. In 1900, the average life span was just thirty-three years; today it is more than seventy-one. The increase has had the most dramatic impact on the world's worst-off. Between 1990 and 2015, the percentage of the world practicing open defecation dropped from 30 to 15 percent. Health inequality has diminished significantly. The world is more literate, child labor has been dropping, we are living in one of the most peaceful times in history. The planet is getting healthier, too. In the past half-century, we have made substantial cuts in indoor air pollution, previously the biggest environmental killer. In 1990, it caused more than 8 percent of deaths; this has almost halved to 4.7 percent, meaning 1.2 million people survive each year who would have died. Higher agricultural yields and changing attitudes to the environment have meant rich countries are increasingly preserving forests and reforesting. And since 1990, 2.6 billion more people gained access to improved water sources, bringing the global total to 91 percent.[13]

Many of these improvements have come about because we have gotten richer, both as individuals and as nations. Over the past thirty years, the average global income per person has almost doubled. That has driven massive cuts in poverty. In 1990, nearly four in ten people on the planet were poor. Today, it is less than one in ten. When we are richer, we live better and longer lives. We live with less indoor air pollution. Governments provide more health care, provide better safety nets, and enact stronger environmental and pollution laws and regulations.[14]

Importantly, progress has not ended. The world has been radically transformed for the better in the last century, and it will continue to improve in the century to come. Analysis by experts shows that we are likely to become much, much better off in the future. Researchers working for the UN suggest that by 2100 average incomes will increase perhaps to 450 percent of today's incomes. Life expectancy will continue to increase, to eighty-two years or possibly beyond a hundred years. As countries and individuals get richer, air pollution will reduce even further.[15]

Climate change will have an overall negative impact on the world, but it will pale in comparison to all of the positive gains we have seen so far, and will continue to see in the century ahead. The best current research shows that the cost of climate change by the end of the century, if we do nothing, will be around 3.6 percent of global GDP. This includes all the negative impacts; not just the increased costs from stronger storms, but also the costs of increased deaths from heat waves and the lost wetlands from rising sea levels. This means that instead of seeing incomes rise to 450 percent by 2100, they might increase "only" to 434 percent. That's clearly a problem. But it's also clearly not a catastrophe. As the UN climate panel put it themselves:[16]

> For most economic sectors, *the impact of climate change will be small relative to the impacts of other drivers* [such as] changes in population, age, income, technology, relative prices, lifestyle, regulation, governance, and many other aspects of socioeconomic development [italics added].[17]

This is the information we should be teaching our children. The young girl holding the sign "I'll die of climate change" will not, in fact, die of climate change. She is very likely to live a longer, more prosperous life than her parents or her grandparents, and be less affected by pollution or poverty.

But because of the fear-mongering surrounding climate change, most people don't hear the good news. And because we believe that climate change is a much bigger challenge than it really is, many countries are spending more and more to combat it, and spending it in less and less sensible ways. Evidence shows that globally we are now spending more than $400 billion annually on climate change, through investments in renewables, in subsidies, and in lost growth.[18]

The costs are likely to keep increasing. With 194 signatories, the 2015 Paris Agreement on climate change, the most expensive pact in human history, is likely to incur costs of some $1–$2 trillion per year by 2030. With ever more nations making promises to go carbon neutral over the next decades, these costs could escalate to tens of trillions of dollars annually in the coming years.[19]

Any response to climate change will cost money (if addressing the problem made money, doing so wouldn't be contentious and we'd already be doing it). If a relatively low-cost policy could fix most of the problem, that could be money well spent. However, it turns out that the Paris Agreement in its best-case scenario will achieve just one percent of what the politicians have promised (keeping temperature rises to 1.5°C [2.7°F]), and at huge cost. It is simply a bad deal for the world.[20]

Moreover, it is unlikely that the Paris Agreement, or any other wildly expensive climate initiatives, will be sustainable. While many people are worried about climate change, most aren't willing to spend much of their own money to solve the problem. Across the world, people are saying they're willing to pay $100–$200 a year to address climate change. A 2019 *Washington Post* survey showed that while more than three-quarters of all Americans think climate change is a crisis or major problem, a majority was unwilling to spend even $24 a year on fixing it. Yet, the commonly proposed policies will cost many thousands or even tens of thousands of dollars per person per year.[21]

When fighting climate change becomes too expensive, people will stop voting for it. Voters have already rebelled against environmental policies that push up the costs of energy: in France this takes the form of the Yellow Vests movement, and in the United States, Brazil, Australia, and the Philippines, it has seen the election of politicians campaigning against climate change policy. For this reason, less grandiose responses to climate change might also be more effective, because the electorate won't turn against them. Climate policy has to be steady to be effective over the long run, and if the costs of climate policy are so high that citizens consistently turn against the governments that promote it, then meaningful change will be hard to come by.

ONE OF THE great ironies of climate change activism today is that many of the movement's most vocal proponents are also horrified by global income inequality. They are blind, however, to the fact that the costs of the policies they demand will be borne disproportionately by the world's poorest. This is because so much of climate change policy boils down to limiting access to cheap energy.

When energy becomes more expensive, we all end up paying more to heat our houses. But because the poor use a larger share of their incomes on energy, a price increase burdens them the most. In the rich world, an estimated two hundred million people already suffer from energy poverty, meaning energy sucks up one-tenth or more of their income. So they either have to use less energy, or they have to cut spending on other things. But energy poverty isn't just an extra cost to the already vulnerable—it can disrupt their lives. For instance, energy poverty means that poorer, elderly people can't afford to keep their homes properly heated, leaving them to stay longer in bed to keep warm. The elite use only a small portion of their large incomes on energy, so even dramatic price increases matter much less to them. This is why it is easier for the rich to argue for high energy taxes. In fact, financial benefits from climate policies (like subsidies given to a homeowner for erecting a solar panel or insulating a house, or driving a Tesla) overwhelmingly go to the richest.[22]

In poor countries, higher energy costs harm efforts to increase prosperity. A solar panel, for instance, can provide electricity for a light at night and a cell phone charge, but it cannot deliver sufficient power for cleaner cooking to avoid indoor air pollution, a refrigerator to keep food fresh, or the machinery needed for agriculture and industry to lift people out of poverty. Countries in the developing world need cheap and reliable energy, for now mostly from fossil fuels, to promote industry and growth. Not surprisingly, a recent study of the consequences of implementing the Paris Agreement showed that it will actually increase poverty.[23]

Our extraordinary focus on climate also means we have less time, money, and attention to spend on other problems. Climate change frequently sucks out the oxygen from almost any other conversation about global challenges. In rich countries, this monomaniacal focus means we have fewer and shorter conversations on how to fix our pension plans, improve our schools, and achieve better health care. For poor countries, climate policy threatens to crowd out the much more important issues of health, education, jobs, and nutrition. These are the issues that, if addressed appropriately, we *know* will help lift the developing world out of poverty and generate a much better future.

SO WHAT IS the way forward?

First, we need to evaluate climate policy in the same way that we evaluate every other policy: in terms of costs and benefits. What that means in this case is that we have to weigh the costs of climate policies against the benefits of fewer climate-related problems. The climate problems are incessantly highlighted, but the costs of a policy for cutting carbon dioxide are just as real, and often hit the poorest in society hardest. Carbon dioxide is a by-product of a society with access to reliable and cheap energy, which helps produce all the things that make it good: food, heating, cooling, transportation, and so on. Restricting access to more costly and/or less reliable energy incurs higher costs that reduce economic growth.

In the case of carbon dioxide, the best research on costs and benefits shows that we should cut some, but by no means all, carbon dioxide emissions. We should do so through a carbon tax, starting out rather low at $20 per ton of emissions (equivalent to an 18¢ per gallon tax on gasoline) and slowly increasing it over the century. The tax should preferably be coordinated globally, but more likely we'll end up with a patchwork of less effective policies. Still, this will cut the global temperature rise somewhat and prevent us from reaching the most damaging temperatures. It will also slightly slow economic growth, because that is the inevitable corollary of making energy more expensive.

Overall, this turns out to be a good deal. We will examine the inner workings of these climate-economic models later, but here is the gist. The cost from slightly more expensive energy translates into a slightly slower-growing global economy that over the next centuries achieves slightly less welfare than it would have without carbon taxes. In short, the extra cost is about 0.4 percent of total GDP.

The lower temperature rise will lead to fewer climate damages over the coming centuries than the world would otherwise have seen. In total, that benefit is worth about 0.8 percent of total GDP. The simple point then is that it is a good deal to pay 0.4 percent of GDP to obtain a benefit of 0.8 percent of GDP.

Cutting some carbon dioxide makes a lot of sense. First, it is easy to cut the first tons, because these are the lowest-hanging fruit. There are many places where efficiency can be obtained at low cost. You can stop heating the patio when nobody is outside, incurring just the minimal inconvenience of turning the heat off. Also, cutting these first tons has the largest benefit, because it cuts the highest and most damaging temperature rises.[24]

But it is also important to recognize the scale of this solution. We pay 0.4 percent and make the world 0.8 percent better off. In total, the benefit is 0.4 percent of total global GDP. Getting a carbon tax right can make the world better, but not by a lot.

An approach informed by cost-benefit analysis also helps show us what we *shouldn't* do. We should not try to eliminate almost all carbon dioxide emissions in just a few short years. Yet, this is what most campaigners clamor for and most politicians profess to want. If we try to do this, the costs could escalate out of hand. Competently done, we would need carbon taxes equivalent to tens or hundreds of dollars per gallon of gasoline in order to effectively prohibit carbon dioxide emissions in short order. This would cost us about 3.4 percent more of total global GDP. Yet, the extra benefits would be much lower at about 1 percent, making the world overall worse off. It would be a bad deal, even if all policies were done competently, and expertly coordinated across all nations and across the century.[25]

It is much more likely that such panicked climate solutions would be done badly and ineffectively, which could make the total costs incredibly large. We would in essence be paying a fantastically high price for little extra benefit. We would truly leave the world much worse off than it need be.

Let's return to the speed limit analogy. No sensible person would argue that we don't need any speed limits, just as no sensible person would argue that we should do nothing in response to climate change. At the same time, nobody argues that we should set the speed limit at three miles per hour, even though it would save thousands of lives, because the financial and personal costs would be too high for us to bear. And so we find a compromise solution somewhere in the range of fifty-five to eighty-five miles per hour. People who worry primarily about safety will argue for speed limits at the lower end, while those who care more about the financial implications of free movement will argue for the higher end. It's a reasonable range for conversation.

By demanding an immediate and dramatic reduction of carbon dioxide levels worldwide, climate activists are essentially arguing for the three-mile-per-hour speed limit. It's a ridiculous demand, at least for anyone who has to get to work in the morning.

Second, we need to look at smarter solutions to climate change. Top climate economists agree that the best way to combat its negative effects is to invest in green innovation. We should be innovating tomorrow's technologies rather than erecting today's inefficient turbines and solar panels. We should explore fusion, fission, water splitting, and more. We can research algae grown on the ocean surface that produces oil. Because the algae converts sunlight and carbon dioxide to oil, burning that oil will not release any new carbon dioxide. Oil algae are far from cost effective now, but researching this and many other solutions is not only cheap but also offers our best opportunity to find real breakthrough technologies.[26]

If we innovate the price of green energy down below that of fossil fuels, everyone will switch—not just rich world countries but also China and India. The models show that each dollar invested in green energy research and development (R&D) will avoid $11 of climate damage. This will be hundreds of times more effective than current climate policies.[27]

Finding the breakthroughs that will power the rest of the twenty-first century could take a decade or it could take four. But we do know that we certainly won't solve the problem with more empty promises and investment in inefficiency. Innovation must be unleashed.

Unfortunately, we are not doing this now. While everyone in principle agrees we should be spending much more on R&D, the fraction of rich countries' GDP *actually* going into R&D has halved since the 1980s. Why? Because putting up inefficient solar panels makes for good photo ops, and it feels like we're doing something—funding eggheads is harder to visualize.[28]

This is one more cost of the relentless alarmism. Since we're so intent on doing something right now, even if it is almost trivial, we neglect to focus on the technological breakthroughs that in the long run could actually allow humanity to move away from fossil fuels.

Third, we need to adapt to changes. The good news is that we have done this for centuries, when we were much poorer and less technologically advanced. We can definitely do this in the future. Take agriculture. As temperatures rise, some wheat varieties might produce less. But farmers will plant other varieties and change crops, while more wheat farming will take place farther north. This is not cost free, but it will significantly reduce the costs of climate change.

Humans have proven themselves to be ingenious masters at adaptation. We can look to Bangladesh, which has massively lowered the death toll from tropical cyclones since the 1970s by investing in smart disaster preparation and better building codes, or to New York City, which learned from tropical storm Sandy and introduced a range of simple measures like storm covers for the subway system.

Fourth, we should research geoengineering, which mimics natural processes to reduce the earth's temperature. When the Mount Pinatubo volcano erupted in 1991, about fifteen million tons of sulfur dioxide were pumped into the stratosphere, forming a slight haze that spread around the globe. By reflecting incoming sunlight this haze cooled the earth's surface by an average of one degree Fahrenheit for eighteen months.

Scientists suggest we could replicate such a volcanic effect and cool the world a lot at a very low cost. It could also cool the world very quickly, in a matter of days or weeks. In that way, geoengineering could provide us with a potential backup policy if, for instance, we find that the West Antarctic ice sheet has started melting precipitously. Standard fossil-fuel-cutting policies will take decades to implement and half a century to have any noticeable climate impact. Only geoengineering can reduce the earth's temperature quickly.

We should not *do* geoengineering yet, because there might be downsides we haven't investigated. But we should research it to find out if it might offer plausible solutions in some cases.

Fifth, and finally, we need to remind ourselves that climate change is not the only global challenge. To most people, it is not the most important issue—it is in fact the *least* important one. A UN global poll of nearly ten million people found climate to be the lowest policy priority, far behind education, health, and nutrition (see figure I.1). People in rich countries, having much better education, health, and nutrition, tend to be more afraid of climate change, but even for Europeans climate rises only to the tenth-highest concern. For the world's poorest, climate is robustly last.[29]

By focusing most of our attention on climate change, we're ignoring other, bigger issues that if addressed could make the world a much better place for billions of people. Expanding immunization and curbing tuberculosis, improving access to modern contraception, ensuring better nutrition and more education, reducing energy poverty—all of these are well within our power and, if we focused on them, could alleviate suffering for huge swaths of the world's population right now.

Moreover, if we invest more in development, it will also make everyone more climate resilient. Making a community more resilient and

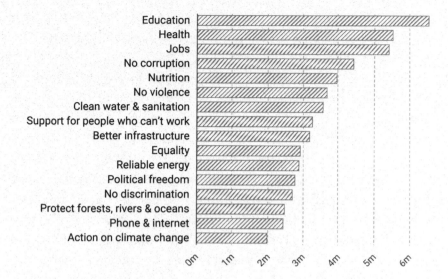

FIGURE I.1 Top policy priorities for the world. In connection with the UN's Sustainable Development Goals, 9.7 million people from across the world ranked their priorities out of sixteen options.[30]

prosperous means more people are able to invest in adaptation and preparedness, and are far less vulnerable to climate shocks. It turns out that helping the extremely poor improve their circumstances also helps them the most with tackling climate.

We need to be aware that when we insist, as part of foreign aid packages, that the developing world align with our climate priorities, we are enacting a kind of imperialism. We are not listening to what the citizens of these countries want. We are jeopardizing their opportunity to lift their populations out of poverty for the sake of our own concerns. This isn't just bad policy. It's grossly unethical.

WE NEED TO take a collective deep breath and understand what climate change is and isn't. It is not like a huge asteroid hurtling toward earth, where we need to stop everything else and mobilize the entire global economy to ward off the end of the world. It is instead a long-term chronic condition like diabetes—a problem that needs attention and focus, but one that we can live with. And while we manage it, we can live our lives and address the many other challenges that ultimately will matter much more for the future.[31]

In this book, we will start by examining the culture of fear created around climate change. Next, we will ask, what does the science actually tell us to expect? What is the cost of rising temperatures? After that, we will look at what's wrong with today's approach. How is it that climate change is at the forefront of our minds, yet we are failing to solve it? What do we achieve by making changes to our lifestyles? What are we achieving collectively, with promises made under the Paris Agreement on climate change? Finally, we will explore how we could actually solve climate change. What policies need to be prioritized in order to rein in temperature rises and leave the planet in the best shape possible for our grandchildren?

We have it within our power to make a better world. But first, we need to calm down.

WHY DO WE GET CLIMATE CHANGE SO WRONG?

PEOPLE ARE PANICKING about climate change in large part because the media and environmental campaigners tell us to, because politicians overhype the likely effects, and because scientific research is often communicated without crucial context. Too often, the missing context is the most obvious fact of all: humans adapt to their changing earth. They have for millennia and will continue to do so. Any projection of the impact of climate change that fails to take this into account is not realistic.

There are strong incentives to tell the scariest possible story about climate change. Media gets more clicks and views with frightening stories. Campaigners get attention and funding. Researchers who position themselves as addressing apocalyptic threats get outsized attention, more recognition for their universities, and more future funding opportunities. Politicians who emphasize the scary scenarios get to promise to save us, and in the process gain the authority to distribute significant resources to fix the problem.

None of this means that we shouldn't worry about potentially big problems. We *want* researchers looking for the big problems, media highlighting what might harm us, and politicians saving us if we need it. But we should be appropriately skeptical, because selling Armageddon is also really useful to all these groups.

We should be most skeptical of the media's coverage of climate change. Nearly every day, we see new stories about rising temperatures and the extreme damage climate change will cause. Again, the media is rewarded for telling the most alarmist possible version of the climate change story—that's what will sell the most newspapers and generate the most clicks. Nobody clicks on a link titled "Life in the future will be very recognizable but could be somewhat more challenging in certain respects." And so instead we read, in the words of one recent headline in the *New York Post*: "Climate Change Could End Human Civilization by 2050: Report." It's highly unlikely that the journalist who wrote the article, or the editor who came up with the shocking headline, was *setting out* deliberately and carefully to mislead their readers. But the journalist and editor most definitely were trying to get more readers. And it's clear that they did not fully read or assess the study they reported on, much less vet it against the established science on the topic.[1]

The actual 2019 study on which the story was based is just a flimsy seven pages from a little-known think tank, and veers wildly from the accepted science of the United Nations climate panel. The report presents the most extreme and unlikely scenario in which all climate impacts are far worse than projected by the vast majority of scientists. Within that extremely artificial setup, which the report authors state is beyond their capacity to model or even quantitatively estimate, there would indeed be a "high likelihood of human civilization coming to an end." But even then, the report does not set the date for the end of civilization at 2050, but only in some unspecified future. As one climate scientist described it: "This is a classic case of a media article over-stating the conclusions and significance of a non-peer reviewed report that itself had already overstated (and indeed misrepresented) peer-reviewed science."[2]

In other words, both the "report" and the news story were more climate fiction than climate news. Yet, in various forms this frightening story made it into *USA Today,* CBS News, and CNN, among many other major media news outlets.[3]

WHAT IS THE media's problem with climate change?

Of course, there is some careful, responsible reporting. But there is much more that isn't. Part of the problem is that over the past several decades, in an effort to seem balanced, many media outlets gave space to climate change deniers long after their arguments had been thoroughly debunked. More recently, deniers are not given space, and this is for the better. But some of the alarmist coverage now may represent an effort to compensate for past sins. Journalists are making the same mistake at the other end of the spectrum: they are failing to hold climate alarmists to account for their exaggerated claims.

Take the June 13, 2019, *Time* magazine cover story. The secretary-general of the United Nations, António Guterres, is photographed standing in a suit and tie with water up to his thighs off the coast of the tiny Pacific island nation Tuvalu. The accompanying article warns that "rising seas threaten to submerge Tuvalu" and states bleakly that because it lies almost at sea level, any rise threatens to wipe Tuvalu and its ten thousand inhabitants "off the map entirely."[4]

Alas, Secretary-General Guterres ruined a perfectly nice suit for no reason: this is not what the science says. Yes, global warming does make sea levels rise, including around the more than 124 reef islands of Tuvalu. But it would have taken the journalists just a few minutes to find the latest scientific study of Tuvalu published in *Nature*. It confirms that not only has the sea level been rising, but around Tuvalu it has risen at *twice* the global average. Yet, during the last four decades of strong sea level rise, Tuvalu has actually expanded and seen its total land area *increase* by 2.9 percent. This is a result of the process of accretion. Yes, sea level rise erodes and reduces land area, but at the same time old coral is broken up by waves and washed up on low-lying shores as additional sand, which counteracts the reduction. The 2018 research shows that this accretion process is overpowering the erosion, leading to net land area gain for Tuvalu. Moreover, this process is ongoing and its dynamic feature will likely mean that the Tuvalu islands can, in the words of the *Nature* study, "persist as sites for habitation over the next century."[5]

The *Time* cover story also warns that two other island nations, Kiribati and the Marshall Islands, will be wiped off the map. A few more minutes of research on both nations would have undercut the entire story. In Kiribati, four atolls all show natural accretion outpacing reduction since 1943. The main Tarawa atoll, where half the population lives, has seen accretion increase the total land area by 3.5 percent over three decades (plus a 15 percent increase from major reclamation projects in South Tarawa). Similarly, the Marshall Islands have seen their total land area increase by 4 percent because of natural accretion.[6]

Indeed, in the latest research summarizing all these studies for Micronesia, the Marshall Islands, Kiribati, French Polynesia, the Maldives, and Tuvalu, it turns out that accretion has beaten out sea level rise on all atolls and all larger islands. Despite sea level rising over recent decades, all atolls studied have increased in area, and all the larger islands studied either remained stable or increased in size.[7]

A more carefully investigated story would have included information on accretion and land mass gain, and could have focused on the challenges facing people who need to move from areas of erosion to accreted land. But instead of looking at the real problems that nations like Tuvalu will face because of climate change, the *Time* magazine story is framed as "our sinking planet": more digestible, scarier, and more saleable. But also deeply misleading.

A SIMILARLY SCARY STORY swept the world in 2019, this one told by the *New York Times* and many, many other media outlets: that vast swathes of inhabited area will be underwater by 2050, with cities "erased." The headlines stem from quality research: a 2019 paper, published in *Nature*, that shows that past estimates of the impact of sea level rise have been wrong, because they relied on measurements of ground level that sometimes accidentally measured the tops of trees or houses rather than the ground itself. This means vulnerability to sea level rise has been underestimated.[8]

That's important. But the media used this to focus on a dystopian vision of 2050. The *New York Times* presented a terrifying map you can see on the left of figure 1.1. The map shows which areas of South Vietnam are under the expected high-tide water line and potentially at risk. Clearly this looks scary, and the newspaper in no uncertain terms declared that it shows South Vietnam will "all but disappear" because it will be "underwater at high tide." It told readers that "more than 20 million people in Vietnam, almost one-quarter of the population, live on land that will be inundated." Similar effects were shown around the world.

This news went viral. "Climate change is shrinking the planet, in the scariest possible way," tweeted Bill McKibben, founder of international climate advocacy organization 350.org. Climate scientist Peter Kalmus

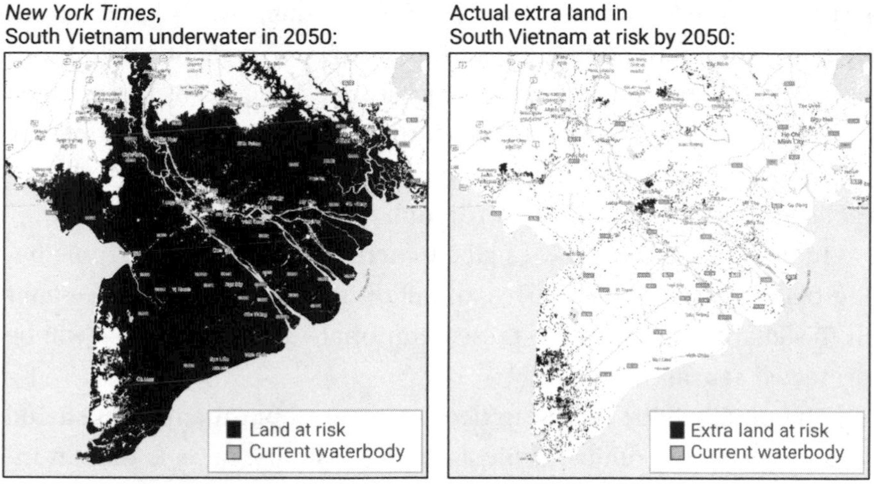

New York Times,
South Vietnam underwater in 2050:

Actual extra land in
South Vietnam at risk by 2050:

■ Land at risk
□ Current waterbody

■ Extra land at risk
□ Current waterbody

FIGURE 1.1 This figure estimates the area of South Vietnam below high-tide water level. At the left, the map the *New York Times* highlighted for readers. The newspaper said it showed that by 2050 this entire part of Vietnam "will be underwater at high tide." This claim ignores existing protection. Indeed, most of South Vietnam is already below high tide, and almost everyone is protected. On the right is the extra land that will be below high tide by 2050. (Left graph reproduced with permission.)[9]

said he was once concerned about being labeled "alarmist," but news like this made him embrace the term.[10]

What did the media forget? To reveal what the comparative situation is today. And it is almost *identical* to the situation estimated for 2050. If you look at the map to the right in figure 1.1, you can see how much *extra* land will be at risk in 2050—almost none. Both maps simply show what everyone knows: people in the Mekong Delta literally live on the water. In South Vietnam's An Giang province, almost all land that is not mountainous is protected by a dike. It is "underwater" in the same way that much of Holland is: large swathes of land including Schiphol, the world's fourteenth-largest airport, are quite literally built under the high-tide mark. In London, almost a million people live below that level. But nobody in Holland, London, or the Mekong Delta needs scuba gear to get around, because humanity has adapted with dikes and flood protection.[11]

The actual research on which the *New York Times* article is based mentions in its introduction that "coastal defenses are not considered" in its approach. That's fine for an academic paper, but it's ludicrous for the media to use its findings to produce claims of "20 million people underwater," or for campaigners to suggest that this gives us reason to all become "alarmist." The study shows that *today*, 110 million people are "underwater" regularly. In reality, almost every one of them is well protected. The real story here is the triumph of ingenuity and adaptation.[12]

In 2050, the study shows a global increase of 40 million people living below the high-tide mark: 150 million in total. As we will see later in this chapter, almost all of these additional vulnerable people will be protected at a fairly low cost.[13]

The media didn't set out to deceive readers, but the news it shared was unnecessarily, unjustifiably alarming. The real news is that an increase of forty million people living below the high-tide mark will be a slight worsening of a challenge that we have shown ourselves completely capable of solving, in a world that will be much wealthier and more resilient than it is today. Context matters.

ONE OF THE most influential recent examples of the media's alarmist approach is its coverage of a major report issued in 2018 by UN climate scientists. Most news outlets reported that these scientists were urging the world to drastically cut emissions by 2030, with huge changes needed to keep temperature rises below 2.7°F (1.5°C). CNN told us, for example, that "Earth has 12 years to avert climate change catastrophe." Versions of this story appeared in newspapers worldwide, and have been parroted by politicians and activists ever since.[14]

In fact, what had happened was that at the Paris climate change conference three years earlier, leaders from around the world had declared that they wanted to achieve the target of keeping temperature rises below 2.7°F. They even put that aim into the preamble of their Paris climate change agreement. They did so at the urging of campaigners who wanted to demonstrate their willpower and ambition, and not because the world's scientists had come together to declare this arbitrary cut-off point crucial.

Having already declared in 2015 that the goal was to restrict temperature rises to less than 2.7°F, world leaders *then* asked the UN's climate scientists to find out what it would actually take to achieve this incredibly ambitious target. The scientists' response became the 2018 report.

The scientists, who have promised to deliver "policy-relevant but not policy-prescriptive information" according to UN guidelines, obligingly said the 2.7°F goal was *technically* feasible but would "require rapid, far-reaching and unprecedented changes in all aspects of society." Simply put, politicians asked them what it would take to do the almost impossible, and the scientists responded that this would require almost impossible policies.[15]

Yet, the report was presented in the media as evidence that we need to make urgent, extreme carbon-emission reductions. It would be a bit like asking the National Aeronautics and Space Administration (NASA) what it would take to move the entire human population to Mars. NASA would obligingly tell us that it is *technically feasible*, but would require far-reaching changes to our current priorities and unprecedented investments in space technology. Campaigners would be similarly wrong in saying, "See, NASA is telling us that we all need to go to Mars."

This dramatic misrepresentation is crucial, because claiming that we have just twelve years left is one of the reasons why children are striking from school, cities and countries are declaring "climate emergencies," and many people are even suggesting we consider suspending democracy to fight this existential threat.[16]

Some people have taken the report to mean that we must rein in temperature rises by 2030, or we will be on a trajectory that will eventually lead to apocalypse. Others take it to mean that climate Armageddon is imminent if the problem hasn't been solved by 2030. Regardless, campaigners and politicians argue that because of the twelve-year "deadline," we should stop even discussing the cost of climate policy: if the world will end, surely nothing else matters.

THE MEDIA'S ROLE in promulgating apocalyptic narratives doesn't fully explain why views about climate change are so extreme on either side. Another important element is the fact that climate change increasingly has become a way to stand out to voters: "I'm going to save you from the end of the world, and *my opponent* won't."

The politics of climate change is increasingly becoming more partisan. In the United States well into the early 1990s, opinion about environmental issues, including climate change, remained remarkably unified. As recently as 2008, former GOP House Speaker Newt Gingrich and future Democratic Speaker Nancy Pelosi filmed an ad for Al Gore's nonprofit organization in which they sat cozily together on a sofa agreeing that climate action shouldn't be partisan.[17]

The era of good feeling has ended. Global warming is now being used, often explicitly, to advance broader causes in a partisan political environment that shapes the United States, the United Kingdom, Australia, and much of the world. This fact goes a long way toward explaining the heightened levels of alarmism that characterize the current conversation about climate change. Up until the 2018 midterm congressional elections, climate change was deemed such a peripheral campaign issue in the US that there was not a single question about it in a general election debate. Things then changed very quickly. By 2019, CNN was hosting an entire "town hall debate" for Democratic presidential contenders framed entirely around the "climate crisis."[18]

A partisan gap in attitudes has been fostered by both sides. Today, people who identify as Democrats and Republicans are further apart on how much priority should be accorded to climate change than on any other single issue. Just consider that. On gun control, the economy, the minimum wage, workers' rights, universal health care, foreign policy, immigration, and abortion, Americans are more aligned than they are on climate change.[19]

Democratic states including New York, California, Washington, New Jersey, New Mexico, Nevada, and Maine have all passed bills requiring "carbon neutrality" by 2050 or earlier. (Being "carbon neutral" means carbon emissions have been reduced to zero, or balanced out by reductions elsewhere.) Republican states have not passed any similar legislation, and in 2019 the Republican Senate minority in Democrat-controlled Oregon blocked a carbon neutrality bill by literally fleeing the state to avoid a quorum. While Democrats have passed what will end up being incredibly expensive promises, President Donald Trump, with Republican support, has done the opposite: he wants to do nothing at all. Neither approach is right.[20]

The partisan divide in America is also reflected globally: overwhelming concern about warming is worn as a badge of honor by leaders of other countries who want to highlight their differences from the Trump administration and its woeful lack of climate policy.

Opposition to Trump has colored coverage of climate policy all over the world. In the wake of Trump's election, for instance, a number of high-profile media outlets began to publish stories claiming that China was stepping up as a "leader" on climate change. A leader on climate change? China has tripled its carbon emissions since 2000 to become the world's largest carbon emitter, and has seen its renewable energy use halve from almost 20 percent in 2000 to about 10 percent in 2020 (although it was even lower at 7.5 percent in 2011). According to official estimates, even if China implements all of its green promises, renewables will reach only 18 percent in 2040, with 76 percent of its energy use still coming from fossil fuels. Holding China up as a green leader is a false narrative that tells us more about the storytellers (and often their opposition to Trump) than it does about China.[21]

SETTING ARTIFICIAL DEADLINES to get more attention is one of the most common tactics of climate change campaigners: if we don't act by such-and-such day, the planet will be doomed. In 2019, Britain's Prince Charles announced that we had just eighteen months left to fix climate change or it would be too late. But this wasn't his first attempt at deadline setting. Ten years earlier he told an audience that he had "calculated that we have just 96 months left to save the world." In 2006, Al Gore estimated that unless drastic measures to reduce greenhouse gases were taken within ten years, the world would reach a point of no return.[22]

But we can go back even further. In 1989, the head of the UN Environment Program declared we had just three years to "win—or lose—the climate struggle." The UN summarized the threat: "We all know that the world faces a threat potentially more catastrophic than any other threat in human history: climate change and global warming." Really? More catastrophic than a potential all-out nuclear exchange? More catastrophic than the one hundred million dead in two world wars in the twentieth century? And more catastrophic than tuberculosis, which in the last two hundred years has killed about a billion people?[23]

Nearly a decade earlier than that, in 1982, the UN was predicting planetary "devastation as complete, as irreversible as any nuclear holocaust" by the year 2000, due to climate change and other challenges including ozone-layer depletion, acid rain, and desertification. And even earlier last century, climate change was causing concern, though for a completely different reason. During the 1970s, while global warming research dominated the scientific community, a number of high-profile researchers promoted fear of a "catastrophic" oncoming ice age. *Science News* had a 1975 cover showing glaciers overwhelming the New York City skyline. *Time* magazine published the story "Another Ice Age?" in 1974, suggesting that "telltale signs are everywhere" for cooling, and that its "effects could be extremely serious, if not catastrophic." Even if there was no ice age, the article told us, just a small drop in temperatures would lead to crop failures, making human life unsustainable.[24]

The fact that we've worried about both cooling and warming does not mean we should not worry about either. The point is that the media likes to predict impending doom, preferably with a firm date attached. And there is something about human psychology that makes us want to believe it.

One of the most striking examples of this apocalyptic tendency came in 1968, when a group of academics, civil servants, and industrialists met in Rome to talk about the seemingly insoluble problems of the modern world. It was a pessimistic age: the techno-optimism of the 1950s and 1960s had given way to concern on a broad range of issues, from geopolitics (the Vietnam War) to society (the "youth rebellion") to the economy (unemployment and stagflation). *Newsweek* summarized the mood with a cover showing a confused Uncle Sam gazing into an empty cornucopia, and the words "Running Out of Everything." In the same year that this "Club of Rome" was forming, the massive bestseller *The Population Bomb* warned that humankind was breeding like rabbits and gobbling up whatever resources it could find, essentially pushing our species "into oblivion."[25]

Against this backdrop, the Club of Rome was determined "to make mankind's predicament more visible, more easy to grasp," as one of the members recalled later. The think-tank membership was convinced that all of humanity was doomed because too many people would consume too much, and we were about to kill ourselves and the planet with over-population, consumption, and pollution. The only hope was to stop economic growth, cut consumption, recycle, force people to have fewer children, and "stabilize" society at a significantly poorer level.[26]

The Club came up with a report, *The Limits to Growth*, which was so influential that it was discussed in magazines from *Time* to *Playboy*, scrutinized by the commentariat, and seized upon by campaigners for radical change. The report had special appeal to the media—and appar-ent extra intellectual heft—because it was based on computer simula-tions, which were then revolutionary and ultramodern. Applying these, the scientists predicted with great confidence that gold would run out by 1979, along with a huge range of important resources that humanity de-pends on—aluminum, copper, lead, mercury, molybdenum, natural gas, oil, silver, tin, tungsten, and zinc would run out before 2004.[27]

Spoiler Alert: They were spectacularly wrong. Consider just the four most important resources. Since 1946, technology has made more cop-per, aluminum, iron, and zinc available than we have consumed, and commodity prices have generally fallen. Oil was supposed to run out in 1990, according to these thinkers and their computer simulations, and natural gas in 1992, but reserves for both are actually larger today than in 1970, although we consume dramatically more of each. Shale gas alone has doubled US potential gas resources within the past six years and halved the price. No resource is infinite. But the resources that can be generated are still far beyond consumption.[28]

The Club of Rome got it wildly wrong because it overlooked the great-est resource of all: human ingenuity at adapting. We don't just use up the iron or gas that is there and then give up. We get better at finding more, at lower cost, in effect allowing humanity access to ever more and ever cheaper resources.

THE STORY OF the Club of Rome is important because lots of people are making exactly the same mistake now when they study and report on climate change: they are leaving out our remarkable ability for *ad-aptation*. Much of the alarmism surrounding the topic can be explained by this one fact: the stories assume that while the climate will change, *nothing else will.*

So, for instance, the *Washington Post* recently reported that "sea-level rise could be even worse than we've been led to expect," swamping an area similar to Western Europe and making 187 million people homeless. Not surprisingly, the notion of 187 million flooded people led many press stories, with Bloomberg News warning that coastal cities around the world were poised to "drown," "swallowed by the rising ocean." Obviously, 187 million makes for a large, attention-grabbing number. Don't believe it. That figure is absurdly exaggerated—and it isn't even new.[29]

The headlines come from a 2019 scholarly paper whose authors simply repeated it from a paper published in 2011. What the earlier paper actually found was that 187 million people could be forced to move in the unlikely event that no one does anything in the next eighty years to adapt to dramatic rises in sea level. In real life, the 2011 paper explained, humans "adapt proactively," and "such adaptation can greatly reduce the possible impacts." When adaptation is taken into account, the authors showed, "the problem of environmental refugees almost disappears." Furthermore, "the main consequence of a large rise in sea level is a larger investment in protection infrastructure" and "it is incorrect to automatically assume a global-scale population displacement owing to a large rise in sea level." Under realistic assumptions, the number of people displaced in an extreme scenario of high sea level rise falls from 187 million to 305,000. The worst-case flooding will displace less than 1/600th of the figure in the headlines.[30]

Journalists and others make this same mistake over and over, with massive consequences for the public's understanding of climate change. In his influential book *The Uninhabitable Earth*, journalist David Wallace-Wells states that coastal flooding caused by sea level rise will result in somewhere between $14 trillion and more than $100 trillion of damages every year by 2100. This idea has been repeated by countless climate activists. But it turns out that these figures exaggerate the problem by up to two thousand times.[31]

Where do the numbers come from? Wallace-Wells uses two key papers to support them. What these papers basically do is predict that sea levels will increase because of climate change over the twenty-first century, and count how many people and how much wealth in those areas will be flooded, without additional flood protection. Do you notice what's wrong with that sentence? Yup, it's the final four words. The headline-grabbing costs come from modeling the effects of flooding *without additional flood protection.*

And "headline-grabbing" isn't an exaggeration. We will come back to the extraordinarily high claim of $100+ trillion, but first let's scrutinize the 2018 research paper that came up with the $14 trillion cost. This was helpfully shared with the world's journalists via a press release. The figure found its way into *Newsweek, Axios, Science Daily, New Scientist,* and *India Today.* What no news stories mentioned, and indeed, what the research paper itself barely acknowledged, was that even an extremely stingy amount of spending on adaptation would reduce costs by 88 percent, and that if we applied real-world, realistic expectations of adaptation spending, the reduction would be far greater.[32]

For climate change to cause $14 trillion of damage, we must assume that not a single country will ever increase the heights of any of its dikes beyond their present levels. They will steadfastly keep their protective walls too low, even as sea levels rise over the century, and even as these countries become much richer (as they will) and able to afford much more protection.

The authors of the original paper acknowledged the lack of logic of this assumption, albeit in fine print: "While the present analysis has focused upon the potential costs of flooding in the absence of additional adaptation from the existing baseline, it is clear that *all coastal nations have, and will continue to adapt* [italics added] by varying degrees to sea level rise." They even point out that "standards of protection are likely to improve particularly with economic growth," making the huge cost of funding even less defensible. Of course, that caveat didn't find its way into the press release.[33]

SOME COUNTRIES HAVE adapted more successfully to climate change than others. Today, coastal cities in the United States have much higher expected damage costs than European coastal cities, because they have much lower protection standards. Likewise, rapidly growing regions in developing countries will likely have a growing adaptation deficit, because coastal development too often takes priority over investments in adaptation.[34]

But there's every reason to believe that globally, adaptation will increase with sea level rise. Studies show that as societies see greater threats, they increase the height and number of protective dikes to reduce these threats. And the evidence also clearly shows that adaptation increases with higher incomes. This makes sense: at the same level of threat, richer countries can afford to demand higher dikes and more protection than poorer countries.[35]

Let's look at the second study that Wallace-Wells relied on to reach the upper end of his estimate, the astonishing figure of $100 trillion or more. The highly quoted study examines the impact of rising sea levels, with human adaptation and without. As you can see in figure 1.2, around the year 2000 about 3.4 million people were flooded each year, the total flood costs were $11 billion per year, and "protection costs" from dikes, levees, and so on, ran to $13 billion per year. The study looks at outcomes based on multiple variables: different levels of sea level rise, population growth, and economic growth. The conclusion is similar across all variations, but let's focus here on what would happen with the highest sea level rises, of almost three feet by the end of the century, in a world with a rich economy that has much to lose.[36]

If we don't adapt, catastrophe ensues. One hundred eighty-seven million people will be flooded each year, and the flood cost will be a phenomenal $55 trillion annually (all in inflation-adjusted dollars). Since we're not spending any more on adaptation, dike costs go up only slightly to $24 billion. In total, flooding costs will make up 5.3 percent of global GDP by 2100, if we don't adapt. In the study's most extreme scenario (not shown in figure 1.2), 350 million people could get flooded every year by 2100, with costs reaching beyond $100 trillion, or 11 percent of global GDP. This is where Wallace-Wells got his high-end, terrifyingly large figure: the worst-case outcome of the worst-case, no-adaptation scenario.

But, of course, we will adapt. As the authors of the paper put it: "Damages of this magnitude are very unlikely to be tolerated by society and adaptation will be widespread." With realistic projections of adaptation,

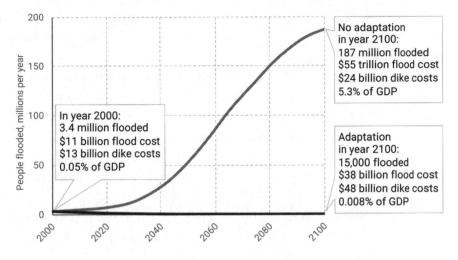

FIGURE 1.2 Number of people flooded over the century with high sea level rise, either ignoring or including adaptation. Cost in percent of GDP includes both flood and dike costs.[37]

the number of people flooded will drop dramatically, to about fifteen thousand per year by the end of the century. Yes, dike costs will increase to $48 billion, and flood damage costs will also increase to $38 billion. But the total cost to the economy will actually *decline*, from 0.05 percent of GDP to 0.008 percent. And a 99.6 percent reduction in flood victims will be an undeniable victory.[38]

Far more than just making it into Wallace-Wells's influential book, this study was quoted by many media outlets, without any mention of adaptation. The narrative that has repeatedly and consistently been told was that 187 million people would be flooded, and trillions of dollars in damages would be caused.

Humans have been adapting to nature for millennia, and with more wealth and technology we will be even better at it in the future. Cutting carbon dioxide is not the only way for mankind to respond to climate change—adaptation is also crucial. For many centuries, much poorer societies have adapted as sea levels have risen. And today we have more know-how and technology than ever before: We can build dikes, surge barriers, and dams; expand beaches and construct dunes; make ecosystem-based barriers like mangrove buffers; improve building codes and construction techniques; and use land planning and hazard mapping to minimize flooding. As a result, deaths from storm surges have actually been declining even as sea levels have risen.[39]

If the fact of ongoing human adaptation to the environment is ignored, the inevitable result is a terrifying story that draws attention. But this portrait is highly deceiving. The reality is that while the amount of carbon dioxide emitted has a comparatively small impact on the number of people flooded, even with the largest carbon emissions and the highest sea level rise, there will be many fewer people flooded because of adaptation, especially in a world that is richer (and the whole world is getting richer and therefore increasingly able to afford it). Even with rising seas, the most likely scenario is that in the future, fewer people will die from climate-related flooding, not more.

WE WITNESS THE SAME sloppy logic on the issue of heat waves. Let's examine a June 2019 headline in *New York* magazine: "Meeting Paris Climate Goals Would Save Thousands of American Lives during Heat Waves: Study." The news story quoted a 2019 study saying that future heat waves so extreme they occur only every thirty years will, by the end of the century, claim a huge number of lives across fifteen US cities.[40]

But here's a very curious thing: that study assumes that nobody in those cities will manage to do anything sensible, like purchasing an air conditioner. For eighty years. So, for instance, the authors project significantly higher death rates over the course of the century in cities like Seattle, where only 34 percent of residents have air conditioners. That some or all of the remaining 66 percent of the population might buy an air conditioner at some point falls beyond the imaginative range of the authors. The reality is that by the end of the century, most people in cities like Seattle will have bought air conditioners and built houses better able to deal with heat. (Indeed, with technological developments it would be logical to assume the air conditioning will be even better than today's.) The city will also likely have invested in social innovations such as "cooling centers" available for poorer people during heat waves—ideas that are already being used in places like Atlanta.[41]

Adaptations like improved standards and expansion of air condition-ing already allowed New York to reduce heat-related deaths by two-thirds between the 1960s and 1990s. France introduced reforms in 2003 that in-cluded making air conditioning mandatory for elderly care homes. As a result, by 2018 heat-related hospitalizations in France were lower than they had been in earlier, cooler years. And Spain cut heat-related deaths between 1980 and 2015, even while average summer temperatures rose almost 2°F (1°C).[42]

So what happens if we account for the fact that people will, in fact, respond as people always have? Well, it turns out that even with much higher temperatures, toward the end of the century total deaths caused by extreme heat could actually *fall* by seventeen thousand across the United States. A headline rewritten for accuracy would read: "Thousands Fewer Americans Will Die Because of Air Conditioning; Paris Treaty Not Relevant to Story." Once the human propensity for adaptation is taken into account, the numbers on climate change start looking a lot less scary. And adaptation should *always* be factored into any climate change study, because humans are *always* adapting.[43]

ONE REASON WHY people are afraid of climate change is that when you watch the news or read the paper, the weather, increasingly, is por-trayed as frightening. Surely climate change is costing us *more* money and lives? What about the hurricanes ravaging coastlines from Florida and Puerto Rico to Samoa? What about the massive floods and terrifying droughts across the world? These disasters seem to be getting worse and worse every year—right?

Wrong. The reality is that these weather events both in number and severity have stayed the same or even declined over the past century, as we will discuss in chapter 3. However, the *cost* of these events is getting much higher, for reasons that have little to do with climate.

A hurricane or flood hitting a sparsely populated Florida in 1900 would have done relatively little damage. Since then, the coastal popu-lation of Florida has increased sixty-seven-fold. Thus, a similar-strength hurricane or flood hitting a densely populated, wealthy Florida in 2020

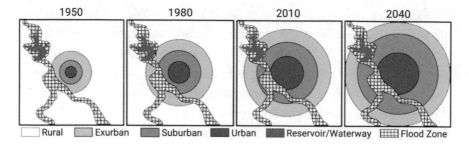

FIGURE 1.3 Illustration of the expanding bull's-eye effect. As a city keeps growing over time, adding more and more expensive housing, an identical flood will have a bigger impact. Thus, the cost of the flooding in 2040 will be much, much higher than in 1950, even though the actual flood event is the same.[44]

leads to much higher costs. The higher cost is not because hurricanes changed, but because society changed.[45]

This is a well-known phenomenon called the "expanding bull's-eye effect": similar climate impacts will result in much more costly disasters because an increasing number of people with more and more valuable assets are at risk. The expanding bull's-eye effect can be thought of as an archery target, with the rings (showing population density) telling us how many people and possessions are at risk of being hit by an imaginary arrow, or natural disaster (see figure 1.3). The rings enlarge over time. This means it becomes ever more likely for an arrow to hit the target—the risk of a huge disaster increases.[46]

Let's look at a real-life example of this: flooding. One 2017 study looked at Atlanta from 1990 to 2010 and found that the number of exposed housing units on its floodplain increased by about 58 percent in just twenty years. This means that with the same amount of flooding and all other things being equal, Atlanta in 2010 on average would have seen 58 percent more housing units flooded than in 1990. Houses in 2010 were also bigger and more valuable than in 1990, so the financial losses would be even higher.[47]

Similarly, a hurricane charging through downtown Miami in 1940 would have wrecked twenty-four thousand homes. The same hurricane hitting today would destroy about a million much more expensive homes. In 2100, it is estimated that a similar hurricane could ruin 3.2 million even more expensive houses.

The expanding bull's-eye effect means we're likely to see much more costly disasters happen over time, even if the climate doesn't change at all. This does not mean that global warming has no distinct impact. But it does mean that when the media trumpets that the latest hurricane, tornado, or flood is the costliest yet, they tend to imply that the rising damage is due to climate change. In reality, much (and often all) we're seeing is that more people with more stuff live in harm's way.

IT IS NO WONDER that people are scared about global warming, given how the media covers it, how campaigners pound it, and how politicians get to grandstand and promise our salvation. Yes, global warming is real, and it needs our serious attention. But the endless jeremiads have warped our understanding of the issues. We need a better sense of what global warming will *actually* mean. Right now, we're getting lots of irresponsible journalism that's scaring us rather than informing us. We need to end the sensationalism and get a grip on the real size of the problem by including information on adaptation and the expanding bull's-eye effect.

As I will argue later in this book, we also need to stop believing that any story with climate in it is best solved through climate policies. As we will see, when we are told about the hundreds of millions of people flooded because of global warming, the argument easily becomes "we need to save them with carbon cuts." But we will find out how little that would help. We will see that even if we went all-in and spent hundreds of trillions on climate policies, *sea levels would still rise, only slightly less* than if we did nothing. Millions would still get flooded. If we instead went all-in on adaptation, we could for less than a hundredth of the cost save almost everyone. The same with heat deaths; focusing on climate policies costs vastly more yet helps much, much less than air conditioning.

Only when the screaming stops will we finally be able to identify the most effective ways to both address global warming and actually help people with their real-world problems.

MEASURING THE FUTURE

WHEN WE THINK about climate change and its impact on our future, we need a clear and understandable measurement system. There are two key variables: temperature and prosperity, commonly measured with "gross domestic product" or GDP. Both are imperfect tools, but they're the best we have.

The concept of climate change includes not just temperature rises but all sorts of other changes, from droughts and floods to storms and reductions in crop performance, from heat-related deaths to rising sea levels. But the summary measure for all these impacts and for climate science in general is the change in global temperature. This doesn't mean that it is the only important yardstick, but it is used as the single most important indicator for all of the potential impacts of climate change.

Similarly, when we talk about human welfare, what we mean can run the gamut from the level of starvation and deaths in society, to access to education and economic opportunities, to the level of happiness and general life satisfaction. GDP does not perfectly encapsulate every one of these aspects of human welfare, but it turns out to be the single measure that more strongly than anything else relates to most of these impacts.

Global temperature and GDP are both rising, and each affects the other. Our efforts to rein in temperatures will cost resources and lead to slower GDP growth. Rising GDP typically means more greenhouse gas emissions, which will speed up rises in temperature. In order to understand what the future holds, we need to get a grasp on exactly what these two variables mean, how they interact, and how much control we have over each.

WE HAVE KNOWN for more than a century that more carbon dioxide in the atmosphere will lead to higher temperatures. The carbon dioxide typically comes from burning coal, oil, and gas, and as the world has dramatically increased its energy use from these fossil fuels, carbon dioxide emissions have kept increasing, almost tripling in the last half century.[1]

Carbon dioxide gas leads to global warming because it lets in the sun's heat but blocks some of the earth's heat from escaping and therefore, a bit like a greenhouse, warms us up. What matters for the earth's temperature is the total amount of carbon dioxide in the atmosphere. Each year's emissions add to the total amount, although the world's oceans and forests suck out the equivalent of about half of these new emissions. So, the amount of carbon dioxide in the atmosphere keeps going up: since 1750, the total amount has increased by 40 percent.[2]

The fact that the climate impact of carbon dioxide depends on all previous emissions emphasizes the size of the problem. Even if we reduce our emissions quite a lot next year, the total amount of carbon dioxide in the air will still go up, just not by as much as it would have. We have to cut a *lot* and over a *long* period to really make an impact. Think of the atmosphere as a bathtub, and carbon dioxide as water. We keep pouring new carbon dioxide into the bathtub, and the plug hole (oceans and forests) allows for about half of the addition to drain away. The total amount of carbon dioxide keeps increasing. Even if we start filling the bathtub more slowly, the total still *increases*, just not as fast. We have to almost stop pouring into the tub for the total amount of carbon dioxide to start decreasing. Even then, it will only start draining very slowly—the plug hole is only so big.

HOW MUCH, AND how quickly, will rising carbon dioxide levels affect global temperature?

Because global warming depends on a lot of factors, scientists use computer models to make projections far into the future. Serious climate models are massive, complex computer programs containing thousands of pages of lines of code. These vast supercomputer models represent the world as little cubes of atmosphere, land, and ocean, and simulate hundreds of thousands of interactions across hundreds of years, including rain, drought, storms, and temperatures for each location while carbon dioxide keeps increasing. These models are expensive to construct and can take weeks or months to run.

But there are also much smaller, faster global warming models that simplify the inputs solely to carbon dioxide and a few other emissions, and model the total change to the global temperature. One of these models is called MAGICC. It was developed partly with funding from the US government's Environmental Protection Agency, and has been used by the UN's panel of climate scientists for all of their reports. We can use the MAGICC model ourselves to calculate the expected temperature rise over this century. For emissions—our input—we will use data from researchers working for the UN, who identified scenarios for the future, ranging from a future in which we remain very dependent on fossil fuels to one in which there is a strong focus on carbon emission reduction. Of these, we will use their "middle of the road" scenario, which fundamentally sees things continuing as they have in the past. Without drastic climate policies, the expectation is that annual emissions will go up and up over the century. This is mostly because the developing world is getting richer, and is expected to continue doing so.[3]

On the left-hand side of figure 2.1, you can see the expected rise in annual emissions over the course of the twenty-first century, more than doubling from today. Plug this data into the MAGICC model, and it will calculate how much temperature will increase as a result. That is what we see on the right: the global average temperature will increase to 7.4°F higher than it was in preindustrial times (the change in temperature would be zero in 1750 if the graph were extended leftward to that year). This model helps us understand what we can and cannot do when it comes to global warming. Let's work out, for example, what would happen if the *entire* rich world stopped all its fossil fuel use in 2020, which would mean grinding its economy to an almost complete halt and leaving it so for the rest of the century.[4]

You can see the impact as the gray line in the left-hand graph in figure 2.2. Emissions drop dramatically in 2020, as the rich world stops emitting one-third of all carbon dioxide. But because most emissions are from poorer countries, emissions keep going up, just at a lower level. In total,

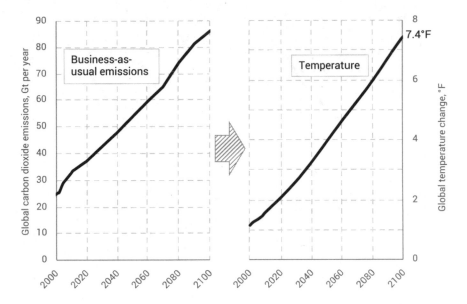

FIGURE 2.1 Emissions and temperature rises. The left shows the UN's middle-of-the-road CO_2 emissions over this century. Put these emissions into the MAGICC climate model, and it outputs the temperature increase across the century, shown on the right.[5]

we would reduce by about a quarter all carbon dioxide emitted over the rest of this century. On the right side of the graph, you can see the effect on temperature over the course of the century. The temperature still rises by just slightly less than it would have. All that matters for temperature is the amount of carbon dioxide in the atmosphere—the contents of the bathtub.

Even if rich countries completely curtail all emissions (an impossible scenario), overall carbon dioxide content continues to rise, and the temperature continues to rise with it. So, the temperature increase is smaller, but only a tiny bit smaller. Even after eight decades, the difference is just below 0.8°F.* Since the United States emits just over 40 percent of rich

---

*We'll use Fahrenheit throughout, because that is how temperature is measured in the United States. But in science (and most of the rest of the world) Celsius (or Centigrade) is used. This only matters because two temperature limits have become focal points for the climate discussion. One is the target of limiting temperature increases to 2°C, which is the equivalent of 3.6°F. This limit was confirmed in the Paris Agreement in 2015, and it is clearly a political limit, not the least because it is a neat and tidy 2.0—but in Fahrenheit, it looks

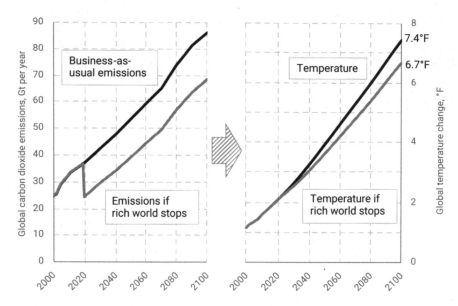

FIGURE 2.2 Emissions and temperature rises with the rich world stopping emissions. On the left, as in figure 2.1, are the UN's middle-of-the-road CO_2 emissions over this century. The gray line shows emissions if the entire rich world ended all carbon dioxide emissions in 2020, and kept them at zero for the next eighty years. When this data is put through the MAGICC climate model, two different temperature outcomes are shown on the right. In a world where the rich countries stop emitting carbon dioxide, the temperature ends up at 6.7°F above what it was in 1750, or almost 0.8°F lower than it otherwise would be by the end of the century.[6]

country carbon dioxide, in this scenario the effect of just the US going to zero fossil fuels from today onward would be a reduction in temperatures of about 0.33°F in 2100.

ON THE ECONOMIC SIDE, let's start by explaining what GDP is: simply put, it is an economy's entire market value of goods and services, and GDP per person divides that figure by the number of people in society.

---

a bit odd, so I'll sometimes write 3.6°F (2°C) to remind you that the Fahrenheit figure relates to the much discussed Celsius target. The other limit is 1.5 °C, or 2.7°F, which has been pushed by many as a more stringent follow-up target to the 3.6°F target. Again its origin is political, but looks better in Celsius.

The measure obviously originates in economics and is constantly referred to in newspapers and by politicians. Yet, some people think using this measurement as a yardstick for well-being is problematic. They say GDP doesn't count intangible things like our health and education, or the joy of a child's play. Measuring welfare in monetary terms, they say, is shortsighted.[7]

GDP doesn't directly measure the health of citizens, but it does include *health spending* on safer births and immunization. It doesn't include the quality of education, but it does measure the higher salaries paid for better teachers, or money spent on computers and textbooks. Higher GDP per person means that the state and people in society have more resources to tackle problems.

On the most basic level, a nation with higher GDP per person is likely to have citizens who live longer; people and the state can afford better health care, nutrition, and safety, and have a host of other advantages that reduce death risks. Higher GDP per person correlates to higher education rates and to lower child mortality because families and communities can afford better education and better health care to treat and prevent disease.[8]

Global growth in GDP per person over the last few decades explains how a billion people were lifted out of poverty. Economic growth has reduced malnutrition by about 50 percent in the last thirty years. And when people become less poor, they get better access to water, sanitation, electricity, and communications technology.[9]

But what does GDP have to do with the planet's health? As countries get richer, and thus their GDP increases, they emit more carbon dioxide. As they move away from agriculture and toward manufacturing, they will be using more and more energy, mostly from fossil fuels, to power their economy, as for instance China has done. And as people get richer, they like their homes better heated and cooled, they build bigger houses, buy more and more varied food, start to travel more, and in general use their income to consume more, which emits more carbon dioxide. Economic growth means that poverty is massively reduced, but at the same time it drives environmental problems like global warming.

But in other ways, rising GDP actually *alleviates* environmental problems, because poverty is often the biggest cause of pollution. One of the deadliest environmental problems today is indoor air pollution, produced almost entirely because the world's poorest 2.8 billion people are forced to cook and heat their houses by burning dirty fuels like wood, dung, and cardboard. Breathing this foul pollution is equivalent to smoking two packs of cigarettes each day, and women and children are the worst affected. When people emerge from poverty, they turn to cleaner gas or electricity as fuel sources. Since 1990, the death risk from indoor air pollution has dropped by 58 percent, mostly because of the increase in GDP per person in the developing world.[10]

The *biggest* environmental killer, outdoor air pollution, initially increases as incomes go up, but then it starts declining as individuals become even richer. Put simply, when immediate concerns like hunger and infectious diseases are tackled, people start demanding more environmental regulations.[11]

Deforestation follows the same pattern. We see vast deforestation in poor countries because there is a strong need for more development, but as countries get richer they become more likely to reforest, in part because citizens increasingly demand more biodiversity and nature.[12]

All of which is to say that we should not assume that rising GDP denotes only bad things for the planet. Higher GDP not only means better social and economic outcomes but also mostly better environmental outcomes. But does money make you happy? Many people say no. Money may solve the problems of the very poorest, according to common wisdom, but beyond a certain level of income, more money doesn't produce more satisfaction. It turns out that the common wisdom is wrong.[13]

If you look across the world (on the left-hand side of figure 2.3), people in richer countries are more satisfied with their lives. This connection doesn't taper off as incomes rise away from absolute destitution: as the national average income doubles and doubles again, the average person is ever more satisfied. And this is true even *inside* each country (on the right-hand side of the figure): as incomes double, individual life satisfaction increases. Even people earning more than half a million dollars annually continue to see life satisfaction and happiness increasing along with their incomes.[14]

Higher GDP per person not only means less death, less poverty and starvation, but also better opportunities, better access to infrastructure,

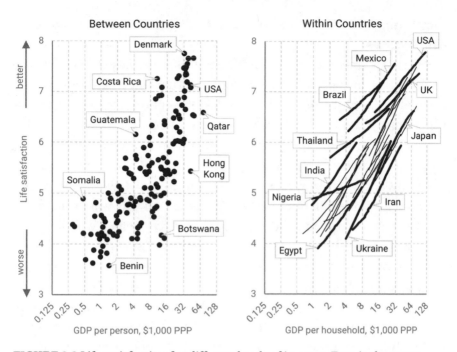

FIGURE 2.3 Life satisfaction for different levels of income. Zero is the worst life possible and 10 is the best life possible. The left graph shows average life satisfaction for each nation. As people get richer, the average satisfaction of the nation is likely to be higher. The right graph shows the same relation inside a nation, for the world's twenty-five most populous countries. Looking at the US line, people at the lowest household income level (less than $16,000 per year) have an average satisfaction level of 6.6; the rich households with $128,000 are on average much more satisfied at nearly 8.[15]

and higher well-being and life satisfaction. That makes it a good yard-stick for human welfare.

OVER THE COMING DECADES, GDP per capita is likely to go up al-most everywhere. It will lift hundreds of millions of people out of poverty and give billions many more opportunities, from avoiding starvation to better education. It will also lead a drive toward better environmental conditions in almost all countries. As incomes go up, voters will press for less air pollution, more protected forests, and cleaner rivers. It will hugely impact global human welfare. It will give almost everyone more satisfied lives.

All this is unequivocally, morally good. But, increased GDP will also drive carbon dioxide emissions higher, and that will create more warming and more problems. And herein lies the central tension within climate policy, and one I will return to again and again. Tackling global warming means limiting the rise in global temperature, or even trying to reverse that rise. This means cutting carbon dioxide emissions dramatically, and as we saw, even stopping the entire rich world's emissions of carbon dioxide will not be nearly enough. And this will mean giving up the cheapest and most reliable energy sources, which will slow GDP growth at significant cost to human welfare.

We have to find the right balance between the two factors we have examined in this chapter. If we focus solely on growing global GDP, we risk temperatures rising to such an extent that the negative impact on our well-being will more than offset the benefits brought about by extra growth. Yet, if we try to cut as much carbon dioxide as we can, out of a sense of panic, we could easily end up reducing human well-being to a degree that far offsets any environmental benefits we achieve.

But if we find the right balance, we can make the world better off overall. We can reduce some of the worst effects of warming while generating sufficient benefits to more than offset the reduction in GDP.

SECTION TWO

THE TRUTH ABOUT CLIMATE CHANGE

A FULLER STORY ON CLIMATE CHANGE

THERE IS A "META-NARRATIVE" about climate change, an overarching storyline that threads through almost everything we read and hear: the headlines of press releases from climate-concerned activists, the anxious sound bites of politicians, the TV news bulletins that pulse with urgency, the movies and books that sketch a bleak future.

The narrative is this: global warming makes things worse, and since it touches almost everything, it makes almost everything worse. The storyline tells us that where there is more rain, the result will be floods and where there is less rain, the result will be drought. Global warming will be good for bad things and bad for good things.

This is a cartoon vision of the world. In real life, most things have both good and bad consequences. You get a new job with a higher salary, which is great, but then the stress causes you to sleep less and eat more unhealthy food. Losing your job feels devastating, but it might also lead you to reevaluate your life and choose a career that you enjoy more.

Like most other things, global warming has both downsides and upsides. More rain can indeed lead to more flooding in some places. But more rain can also alleviate an existing drought. Overall, when we look at all the scientific evidence, while an increase in rain often does alleviate droughts, it generally *doesn't* lead to more flooding, probably because humans use much of the additional water in agriculture and industry.

None of this means that climate change won't in total have significant, negative effects. None of it means that we shouldn't be concerned.

But a one-sided representation means we're not being well informed. We need to see the bigger picture.

According to the conventional narrative about climate change, unless we make dramatic alterations today, animals and people will die in huge numbers, the planet will become unrecognizable, and society will disintegrate. While horrifying, this story is comforting in its simplicity. It is also wrong, mostly because it is a cartoon setup.

In this chapter I will examine just a small handful of the stories that circulate widely about climate change in order to show how the failure to describe and communicate the full picture makes us badly informed and leads to poor policies.

ONE OF THE iconic images of the coming climate apocalypse is the starving polar bear sitting mournfully atop a melting ice floe. Polar bears are adorable, and no one wants them to die. And what better symbol of global warming than endangered polar bears?

In his 2006 hit climate change documentary, *An Inconvenient Truth*, Al Gore showed a sad, animated polar bear floating away on a melting ice floe, presumably to its death. A campaign by environmentalists successfully convinced the US government to declare the polar bear "endangered" in 2008. However, on a global level, the International Union for Conservation of Nature, which decides which animals and plants are endangered, was only willing to call the bear "vulnerable"; this had been the outcome of every evaluation except one since 1982.[1]

The prediction that the polar bear would suffer immensely because of a lack of summer ice was always somewhat odd. Polar bears survived through the last interglacial period 130,000 to 115,000 years ago, when it was significantly warmer than it is now. They also survived the first thousands of years of the current interglacial period, when arctic sea ice cover was strongly reduced and there were even long periods of ice-free summers in the central Arctic Ocean.[2]

When Polar Bear Specialist Group conservationists began studying the polar bear population in the 1960s, they clearly found that the biggest threat was indiscriminate hunting. At that time, it was thought that the global population of polar bears was around 5,000 to 19,000. Hunting

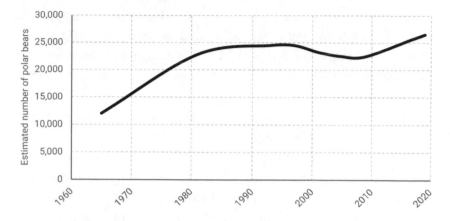

FIGURE 3.1 Estimated number of polar bears from the main international body collecting polar bear statistics, the Polar Bear Research Group.[3]

was regulated, and by 1981 the official estimate had increased to almost 23,000. Since then their numbers have been growing overall. The group's latest official estimate comes from 2019, and is the biggest yet at 26,500 (see figure 3.1).[4]

Clearly, this is a conservation success that we should celebrate. Yet, given that the polar bear has been used for so long as an icon of climate change doom, this finding is actually quite embarrassing for campaigners. The result? Polar bears have been quietly dropped from the climate change narrative.

The *Guardian*, a British newspaper that sees its mission as responding to the "climate crisis," decided in 2019 that it would no longer illustrate as many climate stories with polar bears. (The newspaper went on to announce that it would use more pictures of people in distress in extreme weather events, another problematic choice, as we will see in chapter 4.)[5]

Similarly, the federal government's Arctic Report Card heavily chronicled the decline of polar bears in its 2008, 2009, 2010, and 2014 reports, but now, with official estimates showing that their population is not declining, polar bears no longer get a mention. Nor did Al Gore's follow-up 2017 film, *An Inconvenient Sequel,* find room to share the good news about the bear's survival.

The real threat to polar bears isn't climate change, it's people. Every year around the Arctic, hunters kill almost nine hundred of them. That's more than three polar bears out of every one hundred that exist, every year. If we want to protect them, rather than dramatically reducing carbon emissions to try to tweak temperatures over many decades with a clearly uncertain impact on polar bear populations, our first step should be to stop shooting them.[6]

In fact, when it comes to species extinction, of both fauna and flora, human behavior is a much larger factor than climate change. The World Wildlife Fund Living Planet Report, published in 2018, finds that exploitation (e.g., overfishing) and habitat loss (converting nature into farms and cities) are responsible for about 70 to 80 percent of all threats to species. When we look at what influences species extinction, climate change is one of the smallest factors: 5 to 12 percent. A 2016 study published in *Nature* similarly suggests that overexploitation, agriculture, and urban development are the most prevalent threats to species, with climate change the least important of seven factors. This means that prosaic actions would be most helpful: things like regulating fisheries and ensuring more space for nature. So, yes, if we want to save our plants and animals, we need to change our behavior. But not in the ways that climate activists will tell you.[7]

ASIDE FROM A POLAR BEAR, what's a better symbol of global warming than a heat wave? In recent summers, heat waves in Europe and parts of the United States have become the ultimate evidence that our planet is headed toward an unlivable future. There is nothing good about heat waves—they are dangerous and kill people. But in fact, cold weather is far more dangerous and kills more of the population. This means that as the world gets warmer many people will actually benefit.

Scientists who undertook the biggest ever study of heat and cold deaths, published in *Lancet* in 2015, examined seventy-four million total deaths from all causes in 384 locations in thirteen countries. These included cold countries like Canada, temperate countries like Spain and South Korea, and subtropical and tropical nations like Brazil, Taiwan, and Thailand. The scientists found that heat caused almost 0.5 percent of all deaths, but more than 7 percent of all deaths were caused by cold. For every heat death, seventeen people die from the cold.[8]

Deaths caused by the cold get less attention partly because they are less sudden. Heat kills when body temperature gets too high, and this alters the fluid and electrolytic balance in weaker, often older people. Cold usually kills because the body restricts blood flow to the skin, increasing blood pressure and lowering our defenses against infections.

Essentially, heat kills within a few days, whereas cold kills over weeks. In just the thirteen countries examined in the massive *Lancet* study, one hundred forty thousand people died from the heat each year, and more than two million from the cold. We hear about heat waves that in a few days cause hundreds of deaths, but we do not hear about the thousands of slower cold deaths; this is because there are no TV cameras when mostly elderly, weak individuals expire over weeks or months in anonymous apartments.

Let's take as one example the United Kingdom. Each year, the UK experiences 33 cold deaths for every heat death. In one recent winter in England and Wales, the cold killed 43,000 people. In one year, in a single week in January, 7,200 people were killed who wouldn't otherwise have died. Because of the cold, the mortuaries were overflowing and families had to wait for months to bury their relative. Yet, this didn't become a widely known story because these cold deaths didn't fit the broader climate change narrative.[9]

It is similarly true for India. Recently, CNN capped a monthlong report on the frightening impact of heat in India with the headline "Dozens Dead in One of India's Longest Heat Waves." It was a powerful story told by media and climate campaigners around the world. But it was also rather curious because it focused on the *smallest* risk from temperature-related deaths. In fact, the scientific literature shows that while extreme heat kills 25,000 people each year in India, extreme cold kills twice as many. Indeed, the biggest killer is moderate cold, which kills an astounding 580,000 each year.[10]

CNN could have written plenty of stories telling us "normal cold temperatures in India kill more than half a million people," but it never did.

In chapter 1, we saw that alarming headlines about thousands of future heat deaths are based on a study using the far-fetched assumption that nobody will buy air conditioning as temperatures rise. The latest US data on heat and cold deaths, which unfortunately covers up only to

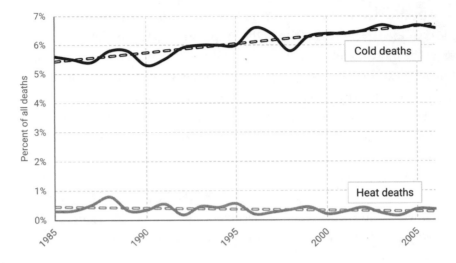

FIGURE 3.2 Percentage of deaths caused by cold and heat in the United States, 1985–2006. Dashed lines show the best linear estimates.[11]

2006, shows that heat deaths are few and are in fact declining (because, in spite of the research assumption, people do buy air conditioning), whereas cold deaths are far larger, and are actually *increasing* (see figure 3.2).[12]

Why? It turns out that reducing death from heat is easier than reducing death from cold, in part because heat is a short-term phenomenon. You need an air conditioner just for a heat wave, and maybe you sit in the room where it's working for the worst days, whereas the cold requires long-term, structural responses such as insulating your home and workplace, and continuous heating throughout the winter months.

If we assume there is no adaptation, it is true that rising temperatures mean heat deaths will increase. But surely, we should also recognize that higher temperatures mean fewer people will die from cold. And since cold deaths outweigh heat deaths almost everywhere, it turns out that even without adaptation, moderate temperature increases will likely mean that the people who *didn't* die of cold will roughly outweigh the additional deaths from heat.[13]

But crucially, it is implausible to disregard adaptation. As temperatures and wealth increase, people will correspondingly make changes to cope. Since it is much easier to adapt to heat than to cold, it is likely that the big killer will remain cold. We don't have global analyses for adaptation to cold and heat deaths, but for the United States, one 2017 study shows that *with* adaptation, even a significant temperature increase will lead to *fewer* deaths overall.[14]

There are two lessons we need to take away. First, hearing only about deaths caused by heat means we end up believing things are much worse, leading to more fear. Second, it means we focus on the smaller problem of heat deaths that in many places is relatively easily solved by simple adaptation measures already being undertaken. Instead, we focus too little on the bigger and often stubborn problem of cold deaths. A 2015 analysis of heat and cold deaths in Madrid showed that not only are cold deaths outweighing heat deaths five to one, but also heat deaths are declining, especially in the older age groups; however, cold deaths are increasing for all groups. The researchers concluded that there is an urgent need "to implement public-health prevention and action measures targeted at cold waves." Even in hot Madrid.[15]

WHAT IS CLIMATE CHANGE doing to nature? Based on what we hear every day, it's turning green pastures and forests into dust bowls. The reality is the opposite. Global warming is causing an unprecedented greening of the world, which scientists have been slow to recognize. But global greening has now been thoroughly corroborated in a number of global studies. The biggest satellite study to date, published in 2016, confirmed that over the past three decades upward of half the world's vegetated area is getting greener, whereas only 4 percent is browning.[16]

The overwhelming cause of global greening is carbon dioxide fertilization. That's right: carbon dioxide, which causes global warming, also helps plants grow—and more carbon dioxide makes them grow more.

Reforestation and more intensive agricultural practices also contribute to the greening of the earth. China has seen its green area almost double in the last seventeen years, partly because of a massive reforestation program, partly because of carbon dioxide, and partly because it grows multiple crops, keeping land green for more of the year.[17]

Researchers find that global greening over the past thirty-five years has increased leaves on plants and trees equivalent in area to two times

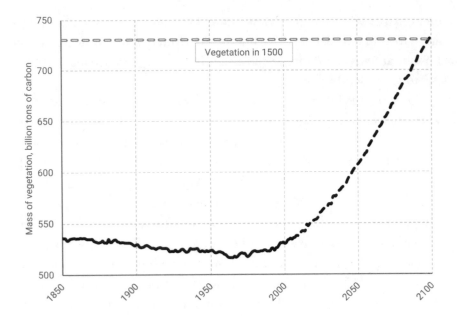

FIGURE 3.3 Weight of vegetation in the world from 1850 to 2100. The black line is an estimate of the actual weight of vegetation, while the dashed line shows what will likely happen if we emit a lot more CO_2 over the century. Estimate for the weight of vegetation in 1500 for reference.[18]

the continental United States. It is equivalent to greening the entire continent of Australia with plants or trees, two times over. It is quite remarkable that over a few decades we got the equivalent of two entire new continents of green because of carbon dioxide—and virtually nobody has heard about it.[19]

As we emit more carbon dioxide over the century, the world will keep getting greener, although there is still a significant debate over *how much* greener. If we measure the weight of all the vegetation in the world, one standard estimate shows that at the turn of the last century we had about 530 billion tons on average over a year (see figure 3.3). The mass of greenery declined slightly over the next seventy years, as people cut down more forest and partly replaced it with farmland containing less green mass.[20]

But something amazing started happening in the 1970s: mostly because of global warming, we started getting more green vegetation. And this will continue over the century as we continue to emit more fertilizing carbon dioxide. Indeed, if we follow a standard worst-case scenario for carbon dioxide emissions to the end of the century, we could end up with almost 50 percent more green mass in the world by 2100. By one estimate, we would be *above* the amount of green the world had in 1500, before we started the widespread reduction in global vegetation.[21]

None of this is to suggest that all is well with global warming—far from it. First, it is possible that not all the vegetation is the type we like or need, although it seems reasonable to say that in general, we would prefer a greener world to a browner one. It also doesn't diminish the fact that many other challenges of global warming remain, including higher sea levels and more heat waves. But it is remarkable how little attention massive global greening has received. It's another way that the global warming discourse is too narrow in scope, focused just on the negative.

THE SPECTER OF more wars is another rallying point of climate change activists. They argue that climate change is a key cause of the ongoing civil war in Syria, and that this war is a harbinger of things to come: warmer temperatures will mean more fighting. The reality is more complicated; in fact, the only robust scientific finding is that warmer temperatures are linked to *less* fighting.[22]

The idea that the atrocities committed in Syria can be blamed on global warming causing a drought is surprising, given Syria's history of bad water management, a population that tripled in thirty-five years and added pressure on resources, the effects of decades of American and British foreign policy in the region, Arab Spring uprisings, religious and ethnic tensions, political repression, and the interference of geopolitical powers. Moreover, recent drought also affected neighboring countries including Israel, Jordan, and Lebanon, yet civil conflict has not occurred on any similar scale there.

A 2014 study of the role of drought and climate change in the Syrian uprisings concluded that "an exaggerated focus on climate change shifts the burden of responsibility for the devastation of Syria's natural resources away from the successive Syrian governments since the 1950s and allows the Assad regime to blame external factors for its own failures." A 2019 study similarly concludes: "There is very little merit to the

'Syria climate conflict thesis.'" Senator Bernie Sanders was knocked by PolitiFact's truth-checkers for making this very argument when he was campaigning for the Democratic presidential nomination in 2015. Yet, the narrative persists.[23]

Moreover, people make a significant but common logical mistake when they pinpoint a specific place like Syria, where global warming could potentially have made the drought worse. In the next chapter we will see that globally, climate has worsened drought in some areas, but reduced it in just as many others. It follows that you cannot make the claim about global warming increasing drought in Syria, and then ignore all the other places where it has *decreased* drought. It's true that over time climate change will likely increase dryness in Syria. But as global warming also increases global precipitation, it will also make many other countries less water stressed, including fragile nations like Guinea, Sierra Leone, Mali, and Burkina Faso. So, if we worry that wars will be worsened in places like Syria, shouldn't we also be thankful that a reduction in future drought from global warming will make civil war *less* likely elsewhere?[24]

On a global scale, the latest survey, published in 2019, shows that "the literature has not detected a robust and general effect linking climate to conflict onset." In fact, studies looking at temperature changes and the onset of wars in the Northern Hemisphere over hundreds or thousands of years show a clear correlation between *colder temperatures* and more war. The reason why cold is a cause of war is that it reduces agricultural production; this leads to social problems, which more often result in war, famine, and population decline.[25]

More importantly, we should be very careful in assigning blame for war to any aspect of the climate. In a recent 2019 study, researchers examined sixteen different factors that drive the risk of conflict. When the scientists ranked that list in terms of influence, climate came in fourteenth, behind far more important factors like poor development, population pressure, and corruption. The scientists concluded: "Other drivers, such as low socioeconomic development and low capabilities of the state, are judged to be substantially more influential, and the mechanisms of climate-conflict linkages remain a key uncertainty."[26]

The narrative that higher temperatures equal more wars is not only facile but ignores the research. The fact is that the only well-documented connection is the exact opposite: over the past thousand years, more *cold* meant more conflict.

WE HEAR AN unrelenting barrage of messages about climate change, and almost all of them are uniformly negative. If that's all we hear, then our understanding of the world is skewed.

Just to be clear, global warming will in total have a negative impact on our planet and well-being, so overall, there will be more bad than good impacts to describe. But the overarching narrative that insists on telling us only the bad stories is unlikely to give us adequate information with which to act smartly. It's important to get the whole picture.

EXTREME WEATHER OR
EXTREME EXAGGERATION?

"A RAGING, HOWLING signal of climate change." That's what the *Washington Post* recently declared about extreme weather. Whether it's flooding, forest fires, or destructive hurricanes, extreme weather events are inevitably cast by the media as evidence that climate change is not only real, but also urgent and disastrous.[1]

But is weather really getting more extreme globally? Are there more and more severe droughts, hurricanes, floods, and fires? To answer these questions, I am going to outline what the science says. As a general rule, I will base findings on what the United Nations' climate scientists—the Intergovernmental Panel on Climate Change, or IPCC—have found. Their reports are generally considered the gold standard, since their work is careful and robust, written by large teams of top scientists from around the world. To see what we should expect just for the United States, I will also bring in official government findings from the US National Climate Assessment.

Many of these findings run against the meta-narrative that everything is getting worse. They seem counterintuitive because we are accustomed to thinking that natural disasters are ever worsening due to climate change. So what really is going on?

Part of the problem is the media. Newspapers fill their pages with stories of bad news because "everything is just fine" isn't a news story. For the past thirty years, polls show, the vast majority of the US population has believed that crime is getting worse, even though statistics repeatedly demonstrate that crime is falling, often quite precipitously. This disconnect can largely be explained by the fact that regardless of how much crime is reduced, there will always be enough to fill newspapers and TV shows every day with horrific stories. Media coverage can help shape a perception that crime is getting more rampant even if statistics show otherwise. Politicians amplify the problem by talking up crime to show themselves "tough on crime," and so statistical reality and public opinion diverge even further.[2]

We are witnessing a similar dynamic when it comes to the effects of climate change. Stories of extreme weather are dramatic—flames engulfing Los Angeles highways, reporters bracing themselves against heavy rains and winds on camera, flood victims rowing through what used to be city streets. Unsurprisingly, people are riveted, and the media dishes up as many of these stories as it can. Global warming is an easy shorthand for explaining these disasters, and many politicians leap in to proclaim themselves "tough on climate change." The scientific facts are left behind because the narrative *feels* true.

You would think that our experience of the weather should be one thing that is completely separate from our political identities. If a Democrat and a Republican look out the window, they should both have the same experience of the weather. Amazingly, that's not true. Democrats, who are much more worried than Republicans about climate change, are far more likely to believe that droughts, floods, wildfires, and hurricanes *where they live* have become more frequent or more intense in the last decade. (And conversely, Republicans are much less likely to believe so.)[3]

It's astonishing to think that people literally living in the same place as you might have a wildly different experience of the weather, based on their political views. But it makes sense: someone who is afraid of climate change will think, "That last storm I experienced felt worse than others: that must be climate change." Similarly, someone relatively unconcerned can always go back in time and think, "Yeah, but the 1935 hurricane was much worse." Both of these sentiments are understandable and underpin the need to ground our approach in more than sentiment alone.

IT'S A CLASSIC IMAGE of climate change: dry, red soil broken up by horrendous drought. Indeed, many arguments for carbon cuts are based on the belief that droughts have already worsened because of climate change.[4]

But in their last major global-level report, the UN's climate scientists didn't find that to be true. They said, "There is low confidence in a global-scale observed trend in drought." They found that since 1950 drought had likely increased in the Mediterranean and West Africa, but had likely decreased in central North America and northwest Australia. In other words, on a planetary level, the earth isn't experiencing more drought.[5]

For the United States, the federal government's most recent National Climate Assessment (NCA) states unequivocally that "drought has decreased over much of the continental United States in association with long-term increases in precipitation." So, contrary to what you hear in the news, even the United States' own climate assessment tells you that drought has *decreased*, not increased.[6]

Moreover, the UN scientists found—and this may be very surprising—that the purported link between man-caused climate change and drought is actually weak: "There is low confidence in attributing changes in drought over global land areas since the mid-20th century to human influence." The US National Climate Assessment agrees. The scientists are basing their findings on data that mostly reveals no increase in global drought. One 2014 study even shows a persistent decline in global drought since 1982, and another from 2018 finds this downward trend goes all the way back to 1902. The evidence also shows that globally, the number of consecutive dry days has been declining for the last ninety years.[7]

For the United States, the National Oceanic and Atmospheric Administration—the official scientific agency that focuses on oceans, waterways, and the atmosphere—finds that since 1895, the area of the lower forty-eight states that has been very dry has certainly not increased. In fact, there has likely been a slight decline, from 12 to 10 percent, although this is not statistically significant. So despite what we often hear, it is incorrect to say that we are seeing the climate impact of drought, either globally or in the United States.[8]

But what about drought in the future? The UN's climate scientists find, with a medium level of confidence, that if carbon emissions increase drastically (indeed, at a level that would be unrealistic), then it is likely that the risk of drought could increase in already dry regions toward the end of the century. Similarly, US government scientists find that if emissions increase at a much faster pace than mainstream scenarios, and if there is no attempt to improve water management, then chronic, long-term drought "is increasingly possible by the end of this century."[9]

So it is possible to argue that climate change can make future drought worse, but it is important to point out the caveats here. This outcome is likely to occur only in scenarios of very high, very unlikely carbon emissions, and the effects will actually be experienced only toward the end of the century. And, as the United States' official report makes clear, this worsening requires an assumption that we wouldn't take any measures to better protect and preserve our water resources.

In the real world, that last assumption is unrealistic. In fact, in California during droughts, reservoirs can be used to reduce the drought deficit by about 50 percent, whereas extensive water usage (mostly irrigation) can almost double drought duration and deficit. Both of these actions, positive and negative, can be more readily, quickly, and efficiently changed than global carbon dioxide levels. Crucially, when we look at the science, the claim that climate change is causing drought today is just not founded.[10]

LEONARDO DICAPRIO'S 2016 climate change documentary was called *Before the Flood. Rolling Stone* published a 2019 article on climate change called "How to Survive a Flooded World." In the same year, the *New York Times* declared: "Flooding Offers a Preview of Future Climate Havoc."[11]

Unlike DiCaprio and the media, when the world's best scientists worked together to examine evidence linking flooding and climate change, they could not find enough proof to even determine whether flooding was getting better or worse. The United Nations has carefully estimated the *total* amount of flooding around the world and found that it's not clear whether it is even getting more or less frequent, much less if there is a human fingerprint involved. Looking at inland flooding, the UN's scientists say there is "a lack of evidence and thus low confidence regarding the sign of trend in the magnitude and/or frequency of floods on a global scale." The US Global Change Research Program clearly says that it cannot attribute changes in flooding to carbon dioxide, nor find detectable changes in flooding magnitude, duration, or frequency.[12]

What's true at a global level clearly isn't necessarily true locally. In some parts of the United States, such as the upper Mississippi River valley, flooding has increased. But it has decreased in other parts of the country such as the Northwest. So overall, the United States' official climatologists have made a similar finding to that of the United Nations: they "have not established a significant connection of increased riverine flooding to human-induced climate change."[13]

Sometimes, people hold up specific examples of flooding (such as in Texas or in Venice, Italy) that are "caused" by climate change, without there being any scientific basis to link the event to climate change whatsoever. Sometimes, there is scientific underpinning to link a flooding event to climate change. What this typically means is that researchers have run a climate model with and without carbon dioxide emissions, and found that running the climate model with carbon dioxide emissions results in more precipitation that could be consistent with the flooding experienced. But we need to stop and think harder about this. If we look at computer models only when there is a flood and then sometimes say a-ha!, we are ignoring all the places where there was no flooding; and, because climate change means less rain in some places, there *could have been flooding* in the absence of climate change. This is exactly what the UN tells us when it points out that overall and *globally* there is no increase in flooding. Reporting only the negative is sadly what gives us a biased understanding.

In the future, the incidence of heavy rain will increase and there will be an expansion of areas that experience significant increases in runoff, which can increase the hazard of flooding, but the UN's scientists emphasize that "trends in floods are strongly influenced by changes in river management." This tells us there are far more important levers we could look at to reduce flooding than carbon cuts. Even in the future, flood damage will be much more strongly affected by other human impacts like river management and by how much building takes place on floodplains, than by climate change.[14]

US government scientists believe future increases in heavy rain could "contribute to increases in local flooding in some catchments or regions" but say it's far from clear *when* any future impact from climate on flooding will be detectable. We know flooding cannot currently be linked to climate, and the US government even points out that we don't know yet *when* or *if* we will be able to claim so in the future. It is hubristic, to say the least, for climate campaigners to claim to know better. Indeed, a 2018 study pointed out that "despite widespread claims by the climate community that if precipitation extremes increase, floods must also," it actually appears that "flood magnitudes are decreasing."[15]

We need to differentiate, of course, between the actual incidence of flooding, which is not clearly increasing, and *damage* from flooding. Images of damaging floods are often used—whether by DiCaprio, Gore, or the world's media—as one of the clear examples of a world already being transformed by climate change. But when we hear the claim that flooding costs are skyrocketing because of climate change, we need to consider the increase in costs resulting from the growing number and value of houses in harm's way, as we saw in chapter 1—the expanding bull's-eye effect.

It is true that inflation-adjusted total flood costs in the United States on average rose from $3.5 billion in 1903 to $12.8 billion in 2018. By 2018, the annual cost of US flooding was 370 percent of what it used to be in 1903. But the number of housing units in the US has increased much, much more: by 2017, there were 750 percent as many housing units to be damaged as there had been in 1903. A flood today will on average make seven and a half houses awash with water, compared to a similar flood in 1903 affecting just one house. Moreover, each of these houses has become much bigger, is worth more, and is filled with many more valuable items. Since 1970 alone, the average house size has increased by half, and the price has almost tripled.[16]

One simple way of adjusting for this dramatic change in the number and value of houses is to compare the flood losses with GDP. This is a weaker correction than if we actually had the value of houses in floodplains since 1903, but it still delivers a very clear result.

Figure 4.1 shows the total cost of US flooding from 1903 to 2018. It shows that in the early part of the last century, when there were few and mostly cheap houses to be flooded, the average flood still claimed about half a percent of GDP each year, with the Great Flood of 1913 costing a

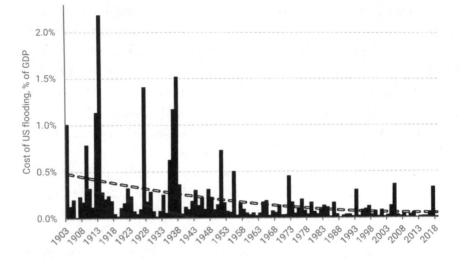

FIGURE 4.1 US flooding costs in percentage of GDP, 1903–2018. Dashed trend shows best fit.[17]

still-unbeaten 2.2 percent. Today, with many more and more costly houses, average floods over a year cost just over 0.05 percent of GDP— ten times less. So while the dollar value of flooding damage has increased dramatically as a percentage of the national income, flooding takes far less of a toll than it did a century ago.

All of which is to say that we need to think about floods differently. Globally, the incidence of flooding is not on the rise, nor is there any evidence that global warming has led to more floods. In some areas, increased rains due to global warming *may* eventually lead to increased flooding, and we should prepare for that scenario with sensible water management policies.

However, we should not confuse the rising costs of flooding with flooding itself (or indeed with climate change). It is entirely caused by more houses and more wealth; in fact, the cost compared to the US national income has declined almost tenfold. If we want to reduce this amount even more, the solution isn't to be found in radically reducing carbon dioxide levels. The solution is to stop building lots of big, expensive houses in flood zones.

EVERY SUMMER IN both the Northern and Southern Hemispheres, wildfires erupt somewhere, contributing to a widely held perception that they are increasing in occurrence, severity, and damage. Perhaps because images of fire are so frightening, or because fire can leap up with apparent randomness, or because (particularly in California) flames have subsumed the homes of the very affluent, wildfires have become one of the most potent political symbols of global temperatures run amok.[18]

The real story bears little resemblance to such alarmist storytelling. To begin with, over the past 150 years our exposure to fire has dropped dramatically. The examination of sedimentary charcoal records spanning six continents and two millennia shows that in fact global burning has declined sharply since 1870. To a large extent, this is because of the so-called pyric transition when humans largely stopped burning wood at home and started burning fossil fuels in power plants and cars. This means that today fire has all but vanished from houses (except those of the world's poorest). By restricting fire to engines and power stations, we have been able to reduce its presence and negative impact in the rest of the world.[19]

There is plenty of evidence for a reduction in the level of devastation caused by fire, with satellites showing a 25 percent reduction globally in burned area just over the past eighteen years. And the primary factor in the reduction in global burned area over the past 110 years is human activity: when more people started planting crops, they wanted to avoid fires, and did so with fire suppression and forest management. In total, the global amount of area burned has declined more than 540,000 square miles, from 1.9 million square miles in the early part of the last century to 1.4 million square miles today.[20]

Recently, scientists undertook a global simulation and found that the area burned for crops and pasture has increased globally since 1900. However, the amount of burning of undisturbed land, and of land that was previously disturbed but is recovering, has declined *more*. Overall, that has had the effect of reducing the total annual burned area by a third.[21]

None of this is to say that wildfires are not a problem. Wildfires in the United States, particularly in California, get a huge amount of global attention. Government scientists undertaking an official 2018 assessment

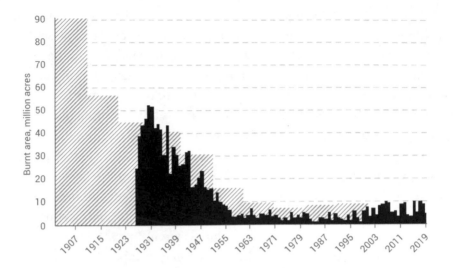

FIGURE 4.2 Wildfire burned area in the United States, 1926–2019, and estimated burned area 1900–2000, broken down by decades.[22]

concluded that the "incidence of large forest fires in the western United States and Alaska has increased since the early 1980s," and they projected these will increase further with higher temperatures.[23]

Indeed, we see an increase in area burned from the three million acres burned on average in the 1980s to the seven million in the 2010s, in figure 4.2. However, this increase is still very small compared to the annual rates of burning we saw in the early part of the last century, with thirty-nine million acres burned on average in the 1930s. So while climate change likely is increasing the amount of land burned by wildfire, and that is something we should take seriously, it does so from a very modest level compared to historical data.

And importantly, it is likely that most of the increase we're seeing now has little to do with the climate and everything to do with other human activities that we have much more control over. A 2017 study found that where humans are present, climate is less important in determining fire activity. It found that significant human presence—such as closeness to towns and roads, the number of people living in an area, and the amount of land developed—can "override, or swamp out, the effect of climate."[24]

Indeed, when we look at the US West, the number of homes built in high-fire-risk zones has increased drastically from half a million in 1940 to almost seven million in 2010. This is more than three times faster than the US-wide housing increase over the same period, so of course many more homes are likely to experience wildfire. Moreover, this growth is set to continue until 2050, meaning our first target for reducing wildfire damage should be to deter people from building houses in high-risk zones.[25]

Damage from wildfire is also on the rise, not because of more wildfire, but because of the expanding bull's-eye effect. There is no US study looking at this, but an Australian study shows that while the value of property damaged by wildfire is on the rise, it is caused by more people and more houses built in high-risk locations. When the damage is adjusted for the number and value of houses at risk, the trend is not increasing—it is actually slightly, but insignificantly, decreasing.[26]

We should take wildfires seriously: global warming will increase the risk of wildfire (though in this century not to the levels mankind experienced prior to the middle of the twentieth century). Compared to the year 2000, an unrealistic worst-case, high-warming trend would increase the burned area globally by 8 percent in 2050 and 33 percent in 2100; but even in 2100 this would still be less than the area burned in 1950.[27]

For high-risk California, global warming by itself will increase the median burned area by 10 to 15 percent by the middle of the century, compared to 2000. But this increase is small in relation to the 50 percent increase in the number of houses in the highest-hazard zone over the same period.[28]

As with flooding, the best way to manage fire is to focus not on carbon dioxide levels, but on human behavior. Planning decisions are far more important than climate impact. Building codes and regulations are of paramount importance, as are land management policies to ensure there are well-maintained firebreaks in forests, and effective firefighting strategies to stop fires from spreading. Above all, if we are serious about reducing future fire damage, regulators and insurance firms need to deliver a harsh but clear message: "You just can't live in areas that are a tinderbox."

HURRICANES, SCIENTIFICALLY KNOWN as tropical cyclones, are the costliest weather catastrophes. The costs of US landfalling hurricanes since 1980 alone amount to two-thirds of entire global catastrophic weather losses over that period. Hurricanes Katrina (2005), Sandy (2012), Harvey (2017), Irma (2017), Florence (2018), and Dorian (2019) have all been used to argue that global warming is making extreme weather worse. But this is not what the peer-reviewed science says.[29]

The UN's climate scientists looked at the evidence and concluded that globally, hurricanes are not getting more frequent: they find "no significant observed trends in global tropical cyclone frequency." They do find an increase in storms in the North Atlantic, but link this to air pollution. They specifically say that there is low confidence in attributing changes in hurricane activity to human influence.[30]

This finding is backed by the US National Climate Assessment, which concludes that hurricane activity in the Atlantic has increased, but that it's not possible to blame climate change. Climate scientists at NASA not only agree, but add that it will not be possible to detect the impact of climate change for at least *a couple of decades*.[31]

Moreover, a new 2018 study reveals that continental US landfalling hurricanes show no trend in frequency or intensity; in fact, to the extent there is one, the trend is slightly (though statistically insignificantly) *declining*. This is true not only for all hurricanes, but also for the worst ones, at category 3 and up.[32]

Yet, the cost of US hurricanes has gone up dramatically. This fact is often used to suggest that global warming is making hurricanes worse and more destructive. But what is really happening is—again—that the bull's-eye is expanding.

While the US population since 1900 has more than quadrupled, coastal populations have increased far more. The population of all the coastal counties from Texas to Virginia on the Gulf of Mexico and Atlantic has increased sixteenfold in the same period. The coastal population of Florida has increased a phenomenal sixty-seven times. There are now many more people living in Dade and Broward Counties in South Florida than lived along the entire coast from Texas to Virginia in 1940. For a hurricane in 1940 to hit the same number of people as a modern hurricane

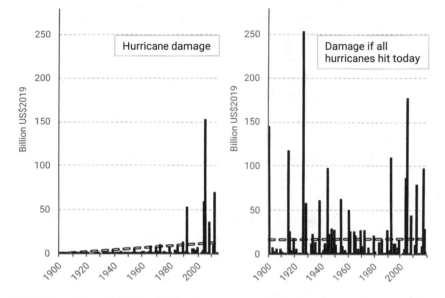

FIGURE 4.3 Cost of US landfalling hurricanes over time. The graph on the left shows the cost of all landfalling hurricanes in the continental United States from 1900 to 2019 in 2019 dollars. The graph on the right shows the cost if the same hurricanes had hit the US as it looks today. Dashed lines are best linear fit.[33]

ripping through Dade and Broward today, it would have had to tear through *the entire Gulf of Mexico and Atlantic coastline.*[34]

From Texas to Maine, the number of housing units within thirty-one miles of the coast increased from 4.4 million in 1940 to 26.6 million in 2000. Correcting for the increase in housing units, the story of worsening damage changes substantially.[35]

If we just look at the inflation-adjusted damage from hurricanes in the left graph of figure 4.3, it is clear that costs have gone up dramatically, something that would still be somewhat true if we look at it as a percentage of GDP, because the increase in coastal vulnerability has increased much, much more than GDP. This is why researchers estimate the cost of hurricanes over time as if all of them hit the United States as it looks today.

The Great Miami Hurricane of 1926 destroyed much of the city (hence the prevalence of Art Deco architecture in the rebuilt areas). Only about a hundred thousand people lived in Miami at the time, in far cheaper houses than today. The inflation-adjusted damage ran to $1.6 billion. A hurricane of the same size and ferocity tearing down the same path today would be the costliest US weather catastrophe ever, causing damage worth $265 billion. Modeling all the two hundred–plus hurricanes that have landed in the United States from 1900 to 2019 as if they landed today corrects for the expanding bull's-eye effect and reveals no significant increase in hurricane-adjusted costs. And this isn't true only for the United States. Similar results are found when scientists look at Australia and China.[36]

Looking to the future, the UN's climate scientists find that the best but weak evidence suggests that hurricanes will become fewer but more intense. These stronger hurricanes will likely create more damage, meaning they will result in more costly damages. But as the population keeps growing and the number of houses close to coastlines increases (it is projected to more than double this century), demographic changes will increase damages much more, swamping the impact of climate change.[37]

The good news is that as people get richer, hurricanes will become less deadly. Poorer communities are affected far more harshly by hurricanes, in part because poorer people live in flimsy housing that offers little protection, and in part because they have little capital or insurance to rebuild. Hurricane Dorian, which hit the Bahamas in 2019, appears to have had its worst effects on the poor Haitian immigrant population whose shantytown, "the Mudd," was devastated.[38]

Over the century, humans will keep getting richer and better able to protect themselves from the effects of hurricanes. Currently, according to a highly quoted study in *Nature*, hurricanes cost humanity about 0.04 percent of global GDP. By 2100, GDP is projected to increase fivefold and deliver a much higher degree of resilience. If we assume that hurricanes stay at the level they are today—that is, with no climate change at all—global hurricane damages in 2100 will cost just 0.01 percent of GDP. However, if we use the UN's expectation of stronger but fewer hurricanes, the global cost in 2100 will double to 0.02 percent of GDP.[39]

So climate change will make future hurricanes more damaging (0.02 percent instead of 0.01 percent), but because the world is getting much richer, and thus more resilient and prepared for disaster, hurricanes will have a *lower* overall cost as a percentage of GDP in 2100 than they do today.

For the future, what matters most is that we make sure that the most vulnerable, worst-off people living in shantytowns like the Mudd are lifted out of poverty. It is growth, not carbon dioxide reductions, that will prevent the harrowing losses that the world's poorest endure as a result of hurricanes.

BECAUSE OF THE expanding bull's-eye effect, the property damage wrought by extreme weather is increasing dramatically. But is extreme weather causing more damage to human life? The answer is a resounding no.

The world's best database of global catastrophes is kept by researchers in Belgium. It includes the number of deaths each year caused by biological means (such as infectious diseases), political catastrophes (such as the early 1930s man-made famine in Soviet Ukraine, called the Holodomor), and earthquakes, volcanoes, floods, and many other natural calamities.[40]

Analyzing this database, we can see that extreme weather used to kill far more people than it does today. Let's look at the climate-related deaths from droughts, floods, storms, wildfires, and extreme temperatures. Since such deaths vary greatly from year to year, we take the average of each decade, starting with the 1920s.

What we see in figure 4.4 is that deaths caused by climate-related disasters have declined precipitously over the past century. In the 1920s, these disasters killed almost half a million people each year, mostly in large floods and droughts in developing countries. Today the total number of climate-related deaths across the world has declined to fewer than twenty thousand each year. Over the past hundred years, the number of deaths from these climate-related catastrophes has plummeted by 96 percent. Remember that over the same time period, the global population has increased fourfold. So the average *personal* risk of dying in a climate-related disaster has declined by 99 percent.

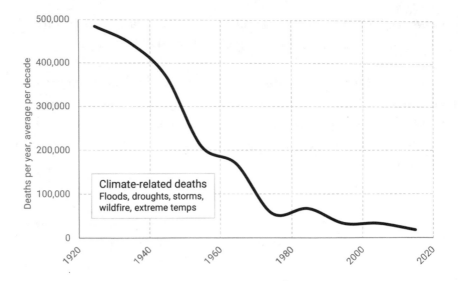

FIGURE 4.4 Global climate-related deaths from disasters, 1920–2019, averaged over decades.[41]

This massive reduction shows a dramatic increase in climate resilience, likely brought about by higher living standards, a reduction in poverty, improvement in warning systems, and an increase in global trade, meaning that droughts are less likely to turn into widespread famines.

Aside from human casualties, the most important impact of disasters is the economic cost: by wiping out a family's livelihood and possessions, a single act of nature can transform and ruin lives. We often hear about "billion-dollar disasters," and how there are ever more of them. Indeed, the *Boston Globe* warns us: "More Billion-dollar US Disasters as World Warms."[42]

And yes, the annual number of disasters costing a billion dollars or more (adjusted for inflation) has indeed increased, from about three in the early 1980s to about fifteen in the late 2010s. But it's once again due to the expanding bull's-eye effect. Any disaster today will cause more damage because there are more homes, factories, office buildings, and infrastructure to destroy.[43]

If we adjust for the increasing size of the economy, a billion-dollar disaster in 1980 would have caused $2.3 billion in costs in 2010, in a 2.3

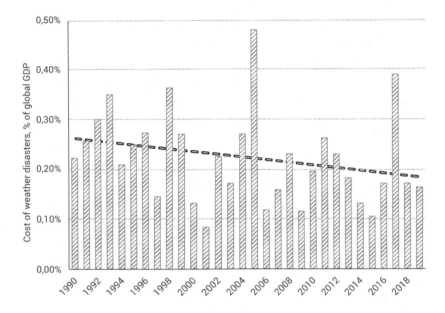

FIGURE 4.5 Global weather-related disaster cost as a share of global GDP, 1990–2019. Dashed line is best linear fit.[44]

times bigger US economy. Once we account for that, we discover that the increase in the number of catastrophic disasters is statistically insignificant.[45]

Globally, we can look at the cost of weather-related disasters adjusted for GDP (see figure 4.5.) We have good global coverage only since the 1990s, but over these past three decades, it is clear that global weather-related costs have not increased, but actually declined, from 0.26 percent of global GDP in 1990 to 0.18 percent in 2019. A new 2019 study splits the damage across all major hazards—from flood, flash flood, and coastal flood, to heat and cold, to drought and wind damage. It finds that for *all* of them, and for both rich and poor countries, the costs of weather-related disasters have declined, in economic terms and, more so, in mortality.[46]

Thus, neither the human nor the relative financial cost of weather-related disasters has actually increased as a result of climate change. We cannot use these numbers to conclude that there are no increases in the number of weather disasters (although as we have just seen, droughts, floods, wildfires, and hurricanes show little or no increase globally), but they do tell us that resiliency has outpaced any potential increase in the incidence of disasters.

And importantly, the total economic impact of these disasters is fairly low. For the United States, hurricanes have cost 0.19 percent of GDP since 2000. Floods have cost 0.07 percent of GDP—that is less than Americans spend each year on fast food. It's a large sum (Americans love fast food), but it's far from a world-ending amount on a national scale. Of course, for an individual or family these events can be utterly devastating—none of what we discuss here is meant to diminish that reality. But a hundred years ago, flooding and hurricane costs were much more devastating for American communities. Both nationally for the United States and across the developed and developing world, extreme weather is causing less suffering both in terms of deaths and in terms of share of GDP.[47]

WHAT IS GLOBAL WARMING GOING TO COST US?

WE NEED TO have a clear idea about what global warming will cost the world, so that we can make sure that we respond commensurately. If it's a vast cost, it makes sense to throw everything we can at reducing it. If it's smaller, we need to make sure that the cure isn't worse than the disease.

Professor William Nordhaus of Yale University was the first (and so far only) climate economist to be awarded the Nobel Prize in economics, in 2018. He wrote one of the first ever papers on the costs of climate change in 1991, and has spent much of his career studying the issue. His studies have helped to inspire what is now a vast body of research.[1]

How do economists like Professor Nordhaus go about estimating the costs of future climate change impacts? They collate all the scientific evidence from a wide range of areas, to estimate the most important and expensive impacts from climate change, including those on agriculture, energy, and forestry, as well as sea-level rises. They input this economic information into computer models; the models are then used to estimate the cost of climate change at different levels of carbon dioxide emissions, temperature, economic development, and adaptation. These models have been tested and peer reviewed over decades to hone their cost estimates.

Many of the models also include the impacts of climate change on water resources, storms, biodiversity, cardiovascular and respiratory diseases, vector-borne diseases (like malaria), diarrhea, and migration. Some even try to include potential catastrophic costs such as those resulting from the Greenland ice sheet melting rapidly. All of which is to

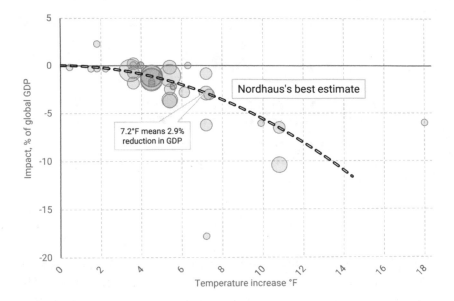

FIGURE 5.1 Impact of temperature rise. Total impact as percentage of global GDP of a given temperature rise, based on thirty-nine published estimates in the literature. Larger circles are better studies. The black line is Nordhaus's best estimate.[2]

say that while any model of the future will be imperfect, these models are very comprehensive.

When we look at the full range of studies addressing this issue, what we find is that the cost of climate change is significant but moderate, in terms of overall global GDP.

Figure 5.1 shows all the relevant climate damage estimates from the latest UN Climate Panel report, updated with the latest studies. On the horizontal axis, we can see a range of temperature increases. Down the vertical axis, we see the impact put into monetary terms: the net effect of all impacts from global warming translated into percentage of global GDP. The impact is typically negative, meaning that global warming will overall be a cost or a problem.

Right now, the planet has experienced a bit less than 2°F global temperature increase since the industrial revolution. This graph shows us that it is not yet clear whether the net global impact from a 2°F change is positive or negative; there are three studies that show a slight negative impact, and one showing a rather large benefit. As the temperature increase grows larger, the impact becomes ever more negative. The dashed line going through the data is Nordhaus's best estimate of the reduction of global GDP for any given temperature rise.[3]

We should focus on the temperature rise of just above 7°F, because that is likely to be what we will see at the end of the century, without any additional climate policies beyond those to which governments have already committed. At 7.2°F in 2100, climate change would cause negative impacts equivalent to a 2.9 percent loss to global GDP.

Remember, of course, that the world will be getting much richer over the course of the century. And that will still be true with climate change—we will still be much richer, but slightly less so than we would have been without global warming.

YOU MAY VERY FAIRLY be asking at this point, how can this possibly be true? Given all the alarming stories in the media, how can it possibly be that the world's top climate economists find that relatively unrestrained warming will cost less than 3 percent of global GDP? It seems like a tiny number given that apocalyptic rhetoric characterizes the climate debate.

We've seen in previous chapters that much of what we have been told about climate change's effects is misleading. Those impacts are the building blocks, as it were, of the total cost. When we allow for the reasonable assumption that people, communities, and countries will take reasonable measures to adapt to the effects of rising temperatures, then the damaging impacts of climate change turn out to be rather low. Indeed, many impacts, as we saw with hurricanes and extreme weather in general, might actually lead to lower, not higher, relative costs.[4]

Let's zoom in on one of the most important areas: agriculture. Over the past century, agricultural output has grown enormously, as farming has become more efficient. Globally, the value of agricultural products is more than thirteenfold higher than 150 years ago. Cereal production is now more than three times what it was in 1961, outpacing population growth that grew a bit more than two times. And these increases are projected to continue.[5]

But global warming will incur real costs and reductions to agriculture. Unlike almost any other sector, agriculture is acutely vulnerable to changes in weather, because most of it will always have to take place outdoors. As rising temperatures and changing precipitation patterns impact future agriculture, the increases in food production will be smaller than they could have been. Reading the headlines, we would think these reductions would be massive: "Climate Change Threatens Future of Farming in Europe," "Climate Change Could Lead to Food Shortages in the U.K.," "Climate Change Is Coming for Australian Sheep," "Climate Change Could Lead to Major Crop Failures in World's Biggest Corn Regions," "Climate Change Is Likely to Devastate the Global Food Supply."[6]

But these headlines are misleading. One broad study done for the UN's Food and Agriculture Organization (FAO) predicts that by midcentury, climate change will reduce global crop output by just a fraction of one percent of today's output. By 2080, in a worst-case scenario, production of cereals (including wheat, rice, maize, barley, oats, sorghum, and quinoa) will be 2.2 percent lower than it would be without climate change. So grain production will still increase overall, just by less than it could have done. The FAO expects an increase in global cereal production of 44 percent without global warming, and this study shows the increase might be reduced to 41 percent by global warming.[7]

So why do the news stories give us the wrong idea? Many of them make two fundamental errors, which distort our impression of the cost.

The first mistake is leaving out the fertilization effect of carbon dioxide. Carbon dioxide is a fertilizer that boosts photosynthesis. That's why professional vegetable growers pump it into the greenhouses that grow tomatoes, cucumbers, and lettuce. More carbon dioxide means bigger, more productive plants. On a planetary scale, this link explains why we are seeing planetary greening.

Many studies leave out the carbon dioxide fertilization effect to make the analysis simpler. This may be relevant in a research context, but when news stories consistently report on these findings without the carbon dioxide fertilization effect, they will dramatically exaggerate the negative effect of climate change on agriculture. One 2018 study, for example, revealed that stringent policies to reduce carbon emissions could increase global crop yields by 22 percent compared to doing nothing. That makes for good headlines. But this is true only if we ignore the effect of carbon dioxide fertilization. The research actually shows that if we include it, the negative effects of higher temperatures are more than counteracted by fertilization, so stringent carbon cuts could really mean 12 percent *lower* crop yields.[8]

The second mistake the alarmist stories make is ignoring the reality of adaptation. For generations, farmers have adapted, and will continue to do so. If you assume that a wheat farmer in today's warm countries will naively and robotically continue to plant wheat even as temperatures rise, then significant reductions in yield are inevitable. But in any real-life scenario, over the next eighty years farmers, their children, and grandchildren will start to tinker with earlier sowing, adapt to different wheat varieties, and might eventually switch entirely to crops that fare better in hotter weather.

At the same time, more wheat farming will take place farther north in cooler areas. One study shows that when researchers ignore such adaptive measures undertaken in the rich world, their yield predictions are fifteen percentage points too pessimistic.[9]

So, when we account for these errors, what will be the total cost of climate change on agriculture? The biggest study showing the total GDP impact is based on 1.7 million fields growing the ten most important crops. The 2016 study uses the differences in these many fields to most accurately assess how changing temperatures will change output and encourage adaptation. It also incorporates carbon dioxide fertilization. Crucially, it includes agricultural trade, which will reduce impacts because more food will be produced in high latitudes and exported to low latitudes.[10]

The full study concludes that the total average global cost will be 0.26 percent of GDP by the end of the century. While we still will end up producing much more food for the world, global warming will mean we have to produce it with more effort and with more trade, overall leaving us less well off, at about a quarter of a percent of global GDP. And actually, this is in a worst-case scenario with very high warming; with less extreme warming the costs end up much closer to zero, and in three of eleven scenarios explored by the researchers there would actually be benefits of up to 0.15 percent of GDP.[11]

This kind of study doesn't give a great, worst-case headline that can bounce around the world. But it does give us a good sense of the size of the total problem.

One of the reasons why the impact of global warming on agriculture is so slight is that agriculture isn't a huge part of the world economy anymore. In the part of today's world that is rich, agriculture makes up an ever smaller proportion of the economy. In the United States, agriculture employed perhaps 80 percent of the workforce in 1800, and it was responsible for more than half the net worth of the economy. Today, it employs 1.3 percent and produces 1 percent of the economy. As countries get rich, an ever smaller proportion of the workforce produces more food, while everyone else is freed to produce other goods and services. Even if we will have to use significantly more resources to grow the same amount of food later in this century, it will be a significant amount of a very small part of the economy.[12]

The poorer world is undergoing the same transition that the rich world has experienced. In 1991, more than half the workforce in the low- and middle-income world was engaged in agriculture and responsible for more than 18 percent of the economy. Today, only a third of the workforce produces much more food, but because the total economy has grown even more, agricultural output accounts for only 8 percent of the economy.[13]

Climate causes only a 0.26 percent reduction in GDP by the end of the century because it affects this ever smaller part of the economy. That serves to underscore how it is possible for the total impact of climate across all areas to add up to less than 3 percent of global GDP, as Nordhaus's and many other studies show.

IT'S NATURAL TO ASK at this point, what about the scary scenarios that the researchers haven't thought of? As we have seen, economists have studied lots of different climate impacts very specifically. But it's sensible to wonder, what if we're still missing something? It's impossible, of course, for any damage model, even the most sophisticated one, to include every single possible climate impact. It stands to reason that sectors most likely to suffer the most expensive damage have been studied

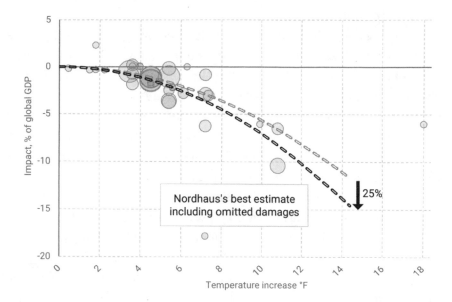

FIGURE 5.2 Impact of temperature rise with unquantified costs accounted for. Same graph as figure 5.1, showing the total impact in percentage of GDP of a given temperature rise. To reflect unquantified costs, the adjusted best estimate has added 25 percent in costs.[14]

and modeled, but there could be omissions that make the estimates too low.

So, to account for any omitted damages, Nordhaus has added 25 percent to each damage estimate to take account of possibly omitted damages (see figure 5.2). While this addition is consistent with those used in other studies, it is still something of a judgment call, since it essentially puts a price tag on what hasn't been analyzed.[15]

Accounting for these additional costs, a temperature increase of 7.2°F around the end of the century would mean a reduction in global GDP of about 4 percent (actually 3.64 percent, but let's just call it 4 percent), as compared to the 2.9 percent reduction projected without the added costs.

HOW TRUSTWORTHY IS the estimate, though? Couldn't there be some big and incredibly costly catastrophe lurking in the shadows?

Probably not.

One possible catastrophe that people often bring up is the complete melting of the Greenland ice sheet. If it happened within a lifetime, that would be catastrophic. However, the UN climate panel finds that even with a very large amount of heat, it will take a millennium or more to melt all of Greenland's ice. Studies indicate that even in the absence of climate policy, 60 to 70 percent of the Greenland ice sheet would be around in a thousand years. And if temperatures went back down over the next centuries, it is likely much of Greenland would regrow. That is why a 2019 study by Professor Nordhaus shows that even in a worst-case, high-temperature scenario, the increased melting of Greenland would have a rather trivial impact by the end of the century, at 0.012 percent of GDP.[16]

Another serious concern often raised is ocean acidification. The cost of ocean acidification is not included in any of the computer models that create estimates of the costs of global warming. The basic problem comes from the earth's oceans taking up carbon dioxide from the atmosphere, making them more acidic. Acidification hurts marine organisms that build their shells and skeletons from calcium carbonate.[17]

Economists have tried to estimate the impact of acidification by exploring what would happen in a worst-case scenario. Here they model a complete collapse of fishing of wild fish from the oceans, both for commercial, recreational, and subsistence fishermen by 2020. Moreover, they imagine a complete loss of tourism and recreation from coral reefs.[18]

This scenario sounds devastating, although it would not spell the end of fish consumption. Remember, two-thirds of the total, global value of fish is already produced in aquaculture, where increasing acidification would have close to zero impact. Yet, the loss of ocean fisheries, recreational and subsistence fisheries, and all coral reef tourism and recreation is clearly significant. The researchers estimate the worst-case cost of this complete collapse at $301 billion by 2200. With middle-of-the-road estimates of economic growth, this translates into a loss of 0.0075 percent of global GDP by then.

We need to bear in mind, as we are using Nordhaus's estimates, that he already added a significant buffer of 25 percent to include any major uncounted costs. This led to our expected global cost increasing from 2.9 percent of GDP to 3.64 percent at 7.2°F, a difference of 0.73 percentage points. It's clear that this increase is much, much higher than the combined century-wide costs of the collapse of the Greenland ice sheet and the total collapse of marine fisheries and coral reef tourism from acidification. And the buffer is large enough to account for many other costs that have potentially been left out. Indeed, this buffer could accommodate a hundred different impacts, each as large as the worst-case complete loss of marine fisheries and coral reef tourism in 2200.

WE HAVE FOLLOWED the UN's overview of the costs of climate change impacts across different temperatures, and used the best estimate from Professor Nordhaus, and we have even added 25 percent to that. This has given us the best available estimate of the total costs of climate change at about 4 percent of GDP by 2100. This estimate is actually larger than the one provided by the 2018 UN climate panel report—the report that everyone uses to argue that we have a deadline of 2030 to act to prevent climate change. This report estimates that if we do nothing, the cost of global warming will reach 2.6 percent of global GDP by 2100.*[19]

When confronted with this economic consensus on the costs of climate change, many climate campaigners often indignantly claim that the "real costs" must be much higher. Yet, when you examine the arguments they use to make this case, they are almost always based on studies or models that leave out adaptation, carbon dioxide fertilization, the expanding bull's-eye effect, and the many other factors we have looked at in this and the preceding chapters.[20]

The economists' work is comprehensive and, in the case of Professor Nordhaus, deserving of the Nobel economics prize. But it is obvious that predicting a cost centuries into the future with exactitude is impossible.

---

*Neither the UN climate panel report nor the Nordhaus estimate fully incorporates the fact that as we get richer, we get less vulnerable. One good example is that as incomes rise to a certain level, malaria gets eradicated because society can afford to invest in serious levels of prevention. Regardless of temperature rises, malaria is simply no longer a threat, as we see in, for example, rich Singapore. This suggests that the 2.6 percent and the 4 percent (if we factor in the 25 percent buffer) are an *over*estimate rather than an underestimate.

What is most important is that while the cost may end up at 3.5, 4.0, or 4.5 percent in 2100, it's unlikely to be 0.01 percent or 45 percent of GDP.

We should take this figure of 4 percent, then, as a guide, which is exactly what we will do in the chapters coming up, where we contrast the cost of climate change with the cost of climate change policy.

HOW NOT TO FIX CLIMATE CHANGE

YOU CAN'T FIX CLIMATE CHANGE

AROUND THE PLANET, well-meaning people are making a myriad of lifestyle changes to reduce their carbon dioxide emissions. Individuals and companies are spending hundreds of millions of dollars annually on "carbon offsets" and other measures. Taxpayers are, through regulation and subsidies, collectively spending hundreds of billions of dollars annually to encourage the use of today's alternative energy technologies, such as electric cars and solar panels. Soon, we will start spending trillions or even tens of trillions of dollars every year in an effort to convert the modern economy to green energy sources.

Yet, all of these efforts are failing. If climate policies worked, then the amount of carbon dioxide we emit for every unit of energy produced, so-called carbon intensity, should be declining. But it is not.

Despite dozens of climate summits, and despite global climate agreements struck in Kyoto and Paris, carbon intensity has *increased* ever since nations first made commitments to rein in climate change at the 1992 Earth Summit in Rio de Janeiro. Carbon intensity is at a higher level than ever before.[1]

Not only does each unit of energy emit ever more carbon dioxide, but the world also uses more and more energy. As a result, total carbon emissions keep rising. Since 1992, humanity has emitted more carbon dioxide than in all history before then. Emissions will likely keep increasing in the coming decades, as more poor countries clamber out of poverty.[2]

The last decade has seen more focus on climate change than ever before. Yet despite this, we are not achieving anything. In a surprisingly honest review of climate policies, the United Nations revealed that the last decade of climate policies has achieved a sum total of nothing.[3]

It's clear that the current approach to climate change is not working. But before talking about what we should do instead, we first need to understand why we're failing. Over the next five chapters, we will explore why our current efforts achieve so little. We'll begin with the first great myth of climate activism: that individuals can make a significant difference.

IF YOU'RE CONCERNED about global warming—and really, who isn't?—you know that you should recycle, eat less meat (or maybe none at all), take public transportation more and drive less, and carry your groceries in recyclable bags. As *Blue Planet* host David Attenborough puts it, every single one of the planet's seven billion inhabitants "must play their part" against global warming and take "simple everyday actions."[4]

We live in a world where Prince Harry and Meghan, duchess of Sussex, are castigated for flying on private planes while Swedish activist Greta Thunberg is hailed for taking a wind- and solar-powered boat from Europe to New York to attend a climate conference. (She was later subjected to sneering headlines when it turned out the trip increased carbon emissions because crew needed to fly to New York to sail it back.) The personal has become highly political. People are worried about climate change, and they want to help solve the problem. The intentions are noble ones. The problem is that the changes we can make to our personal lifestyle and habits at best make only a tiny difference.

Asked what personal action he would take to prevent global warming, David Attenborough once promised to unplug his mobile phone charger when not in use. The logic makes sense: he would save on electricity, which is mostly fossil fuel powered. But if he remembers to do this consistently throughout the entire year, he will cut just seven pounds of CO_2 each year, less than half of one-thousandth of the average emissions for a single person from the United Kingdom. And if the goal is to reduce the emissions caused by a mobile phone, focusing on this trivial personal action means we miss the bigger picture. Charging makes up less than one percent of a phone's energy needs. The other

99 percent comes from manufacturing the phone and operating data centers and cell towers.[5]

Seven pounds of carbon dioxide per year is hard to comprehend. One simple way that we can get a sense of the scale of the effects of different cuts is by using the carbon trading system. The RGGI, or "Reggie," is a carbon-trading marketplace covering the northeastern United States. It is just one of many around the world, but the first and biggest in the US.[6]

The RGGI puts a cap on the amount of carbon dioxide that large, fossil-fuel-run power plants can emit in the region. Then it allows emittance authorizations to be bought and sold. It costs about $6 to buy an authorization for one ton of carbon dioxide. If you buy an authorization, it means there's one ton less that power plants can purchase. If you don't use the authorization, that means all the power plants, between them, need to find a way to emit one ton less over the next year. Essentially, you've spent $6 and reduced global emissions by one ton.

The RGGI puts the actions of individuals in perspective. It gives us a sense, for instance, of the scale of the impact of Attenborough's promise. Cutting seven¢ of carbon dioxide would cost a little less than 2¢ on the RGGI. Attenborough might as well have just donated 2¢ to the climate cause.

SADLY, THE VAST MAJORITY of the actions individuals can take in service of reducing emissions, and certainly all those that are achievable without entirely disrupting everyday life, will make little practical difference. That's true even if *all of us do them*. The UK's former chief climate science advisor, the late David MacKay, once wrote of carbon-cutting efforts: "Don't be distracted by the myth that 'every little helps.' *If everyone does a little, we'll achieve only a little*."[7]

Three central challenges emerge when we try to cut our personal emissions. The first, as we saw in the case of Attenborough's mobile phone, is that cuts are typically small. The second is that we almost always save money. This is a problem because of a well-established phenomenon called the "rebound effect." When we save some cash by being more efficient, we spend the savings elsewhere in ways that lead to more emissions.

Let's take the example of reducing food waste. In theory, if we reduce the amount of food we buy, less food will need to be produced, which will reduce agricultural emissions. But it will also have the happy effect of saving us money. The problem is that we will spend that money on other things, many of which will produce emissions, like an extra vacation. In one 2018 study, Norwegian researchers found that realistically, the money saved from cutting food waste will be spent on other goods that will emit so much carbon dioxide that the original emissions savings will be entirely canceled.[8]

In many cases, the rebound effect doesn't swamp the entire effort; car pooling, for example, will actually reduce emissions, with the cost savings leading only to 32 percent of the reduction being lost. Sometimes, however, the rebound effect leaves us worse off overall. For instance, walking instead of taking the train means we emit much *more* carbon dioxide, because trains aren't big emitters, and we save a lot of money that we spend on other things. Generally, when researchers have studied the rebound effect across a range of activities, they estimate that 59 percent of the emissions savings from "virtuous" behavior are lost to the rebound effect.[9]

A third problem with restricting our behavior for environmental reasons is that as in many areas of life, when we do something "good," we allow ourselves to do something "bad" as a reward. This tendency is known as "moral licensing." This pattern may be familiar to anyone who has struggled with dieting. If you're making good progress on your diet, you're much more likely to choose a chocolate bar over an apple as a snack than someone who is struggling with their diet. This effect is copiously documented when it comes to environmentally friendly behavior. People who have just donated to a charity are less likely to behave in an environmentally friendly way afterward. Those who have reduced their water consumption through an awareness campaign use more electricity instead. One study of shopping behavior shows that the more consumers purchase energy-saving lightbulbs, use eco-bags, or reuse their own bags, the more likely their weekly shopping is to contain meat and bottled water.[10]

I was once invited to participate in a debate with politicians and a journalist whom the BBC dubbed "Ethical Man." This man had just spent twelve months documenting how he and his family had cut their carbon dioxide emissions. He insulated the house, sold the car, cut out meat, and even looked into ecological burials (although nobody died). In total, his family managed to cut emissions by around 20 percent, at a high personal and financial cost. What was most striking to me about Ethical Man was that when he finished his year's hard slog, to celebrate he purchased tickets to take the whole family to South America, blowing his family's entire carbon dioxide savings.[11]

VEGETARIANISM HAS BECOME a major subject in the climate wars. The idea that if you care about the environment you should eat less meat has become commonplace in the West, and vegetarianism is on the rise. Christiana Figueres, a global climate campaigner and former leader of the UN's climate change body, even suggested: "How about restaurants in 10–15 years start treating carnivores in the same way that smokers are treated? If they want to eat meat, they can do it outside the restaurant."[12]

I've been a vegetarian for ethical reasons since I was eleven, and I believe everyone should make up their own minds about whether or not they eat meat. But we should be honest about what it will achieve. That's especially true because going vegetarian is actually quite difficult; one large US survey shows that 84 percent of people who decide to become vegetarians end up failing, with most failing within a year.[13]

At a global level, of course, exhorting everyone to become vegetarian is callously ethnocentric. Right now, the world has 1.5 billion vegetarians, but only 75 million like me are vegetarians by choice. Most are vegetarians because they just can't afford meat, and as they move out of poverty, they will likely eat more and more meat.[14]

But if you actually stopped eating meat, how much would you achieve? Campaigners for "climate-friendly diets" rely on cherry-picked factoids to make the effects sound impressive. Many credulous news outlets suggest that eliminating meat from your diet can reduce your personal carbon emissions by 50 percent or more. That's massive, but also massively misleading. This level of reduction is achieved only by people who go entirely vegan. That means completely avoiding all animal products, including not just meat but milk, eggs, honey, poultry, seafood, fur, leather, wool, and gelatin. It is suggested by other news outlets that vegetarians achieve about half this figure.[15]

Regardless, the numbers remain massive exaggerations: vegans don't eliminate 50 percent of their personal emissions, and vegetarians don't cut 20 to 35 percent of theirs. They cut only that percentage of their *food-related* emissions. And food-related emissions represent only a small fraction of an individual's total emissions.[16]

A thorough, systematic analysis shows that eliminating meat from your diet will reduce your personal emissions by the equivalent of 1,191 pounds of carbon dioxide per year. For the average person in the industrialized world, that means an emissions cut of just 4.3 percent.[17]

We have gone from claims of a 50 percent cut in emissions all the way to less than one-tenth of that. But there's more, because we need to allow for the rebound effect. Vegetarian diets are slightly cheaper: in the United States, vegetarians save about 7 percent, in Sweden about 10 percent, and in the UK about 15 percent of their food budgets. Spending that extra money on other goods and services means the rebound effect likely cuts around half of the saved carbon emissions accrued from going vegetarian.[18]

So if you're living in a rich country, going entirely vegetarian for the rest of your life will reduce your total personal emissions by about 2 percent. You could achieve a similar emissions reduction by eating anything you want and paying $1.50 each year on the RGGI trading system.[19]

Trying to save the world by giving up meat or unplugging our cell phones is ultimately misguided, and it distracts us from doing things that could make a real difference. If we want to improve the carbon efficiency of our food supply, one obvious approach would be to spend more on research and development for artificial meat. If this could be as good as the real stuff (some would say we're already there) there are huge opportunities to cut emissions. Artificial meat generates up to 96 percent fewer greenhouse gases than conventionally produced meat. A perfect meat substitute would be an obvious win because it doesn't require people to give up something they like. People could continue to enjoy their "meat," but with just 4 percent of the emissions. And I would be able to eat burgers again.[20]

ELECTRIC CARS ARE marketed as another great way to reduce carbon emissions. But most drivers are reluctant to switch because electric cars are substantially more expensive and the need for recharging leads to "range anxiety"—"Can I make it to the next charging station?" That is why across the United States and around the world, lavish subsidies are the norm. When governments removed this support, for instance in Denmark and Hong Kong, sales plummeted to nearly nothing.

Replacing your current car with an electric car won't lead to dramatic emissions cuts, alas. The International Energy Agency makes a comparison between a standard gasoline-powered car and an electric one. The gasoline-powered car emits thirty-four tons of carbon dioxide over its ten-year lifetime, including production and disposal. At first glance, an electric car would seem to eliminate all the emissions. They are certainly sold as having "zero emissions," but that is true only while it's being driven.[21]

Yet, in many parts of the world, electric cars are reliant on electricity largely produced from fossil fuels. And their production is also actually *more* energy intensive than that of a gasoline-powered car, especially the battery. Since this production is also typically reliant on fossil fuels, an electric car is actually responsible for lots of carbon dioxide emissions over its lifetime. Across the world, an electric car with a reasonably long range will on average emit twenty-six tons over its lifetime. So, switching from a gasoline-powered car emitting thirty-four tons of carbon dioxide to a comparable electric car that emits twenty-six tons doesn't eliminate emissions; it cuts them by 24 percent, leaving more than three-quarters in place.[22]

Emissions, of course, aren't the only cause of damage caused by driving. Driving creates lots of problems, ranging from noise pollution to lethal accidents. Figure 6.1 shows the European Union's latest estimate of the cost of all damages caused by driving, which is calculated to be almost 20¢ for every mile driven.

Notice that climate damages are a small part of the total damages. Of course, the electric car will be a bit better than a standard car on this measure. Electric cars also make less noise, which is about 5 percent of the total damage caused by driving, but nevertheless a measure on which electric cars outperform standard ones.

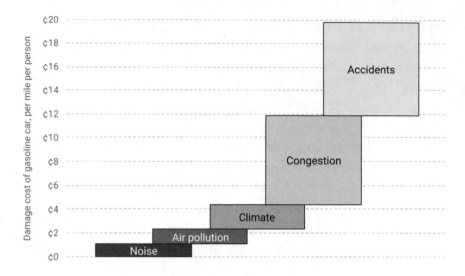

FIGURE 6.1 Damage costs of driving a gasoline-powered car one mile, per person. (Notice the climate impact is likely vastly exaggerated.)[23]

Air pollution has an impact of 6 percent and, surprisingly, more electric cars can mean more air pollution. If you have plenty of hydropower as in Norway, replacing gasoline with electric power can improve air quality, but in places with significant coal-powered electricity generation, including parts of the United States, more electric cars mean *more* air pollution.[24]

China, the world's biggest electric car market, has so many coal-powered plants that electric cars worsen local air, with lethal consequences. It is estimated that in Shanghai, pollution from an additional million electric-powered vehicles would kill nearly three times as many people annually as an additional million gasoline-powered cars.[25]

But pollution isn't the most significant cost when it comes to cars. About 80 percent of the damage cars cause overall comes from accidents and congestion. For these effects, it makes no difference if a driver is in a Tesla or a BMW. All of which is to say that buying, or subsidizing, electric cars doesn't help with the biggest social problems with cars, and it is certainly not a good investment from a climate standpoint. Look at the costs we pay. The average subsidy spent on electric cars globally is about

$10,000 per car. Each car saves eight tons of carbon dioxide over its lifetime, a reduction you could buy on the RGGI market for just $48.[26]

And the impact will be minimal. The International Energy Agency hopes we can reach 130 million electric cars by 2030, a breathtaking expectation given that we have spent decades and billions of dollars in subsidies to reach just 5 million. Even if we could do that, it would cut a trifling 0.4 percent of global emissions by 2030. Electric cars will be part of our future solution to transport needs, but they are not about to solve climate change.[27]

AIRPLANES HAVE BECOME another battleground in the climate wars. Because of the large carbon footprint of air travel, environmental groups are increasingly pushing to make us feel "flight shame"—guilt for the carbon emissions produced anytime we fly.[28]

This is a troubling movement. There has been a tremendous democratization of air travel in recent decades. Whereas flying was for many years the preserve of the ultrarich, now most people in developed countries have the opportunity to travel. But the vast majority of people on the planet have still never taken a single flight—at least 80 percent, according to Boeing. In India, where some 150 million people live in poverty, only 2 percent of the population has ever boarded an aircraft.[29]

Flights are not only for leisure; they allow the flow of expertise and labor, let us solve medical emergencies, provide disaster relief, help us to learn about other cultures, and connect us with loved ones. Even if we were willing to sacrifice all of these things, staying off airplanes would not have the dramatic impact on climate change we might imagine. Even if every single one of the 4.5 billion people getting on any flight this year stayed on the ground, and the same happened every year until 2100, the rise in temperatures would be reduced by just 0.05°F, equivalent to delaying climate change by less than one year by 2100.[30]

What's more, at a personal level there's a huge rebound effect. Researchers found that about 22 percent of the climate change benefits from canceling a personal holiday flight are undone, because the saved money is spent on activities that create emissions. And the rebound effect from canceling a business flight is 159 percent, meaning that the climate impact actually *increases* with each canceled flight. (The rebound effect has much more dramatic implications for a business-class seat because the ticket is so much more expensive, while the carbon footprint is only slightly higher than in the cheaper cabins.) Thinking of saving on emissions by taking a cruise? Sorry, there is evidence that cruise ships are more environmentally damaging than planes.[31]

Instead of telling people not to fly on airplanes, we should focus on the carbon efficiency of those airplanes. Adaptation is already occurring: each new generation of aircraft is on average 20 percent more fuel efficient than the model it replaces. Over the next decade, airlines will invest $1.3 trillion in new planes.[32]

Research is also under way on creating more sustainable fuels, including those produced by domestic and industrial waste. The overall carbon footprint of sustainable fuels is up to 80 percent less than that of today's aviation fuel, and testing started on using such fuels on commercial aircraft in 2008.[33]

Billions are being spent, too, on exploring technologies like more aerodynamic wings, advanced and lighter plane structures, more efficient engines, and new aircraft configurations. And the International Air Transport Association, which represents airlines, estimates that better route management could cut CO_2 emissions by 10 percent. Every dollar we can spend as a society on research and development to bring forward the moment of carbon-dioxide-neutral flights will be far more meaningful and more effective in addressing climate change than a few of us trying to cut a few flights out of a misplaced sense of shame.[34]

WHAT ELSE SHOULD WE give up for the sake of the environment? Children, of course!

A chorus of campaigners, scientists, and journalists have suggested that for the sake of Mother Earth, people should stop reproducing. The *Guardian* directs readers: "Want to Fight Climate Change? Have Fewer Children." The *New York Times* warns that having a child is the worst environmental action possible. A hypothetical American woman might switch to a fuel-efficient car, drive less, recycle, install efficient lightbulbs and energy-saving windows, but the newspaper says that by having two children she will wreak damage "nearly 40 times what she had saved by those actions."[35]

Such exhortations are not new. In the 1970s, American activists founded the National Organization for Non-Parents, promoting child-lessness as a "politically responsible" choice, because they believed the world was collapsing due to environmental ruin.[36]

I think it's a morally bankrupt and intrusive argument; environmentalists have no place telling people that they should not become parents. And what's more, their counting doesn't even add up.

The problem with research on the climate consequences of having children is that it's based on a deeply bizarre way of measuring these consequences. The most cited research paper, published in 2009, measures the climate impact of children by holding each parent responsible for half of every child's projected emissions over the entire duration of the child's life. Fine so far. But they do not stop there. A parent is held responsible not just for half of their child's emissions, but also for a quarter of their grandchildren's emissions, an eighth of their great-grandchildren's emissions, and so on. Moreover, the researchers estimate that each of these future children will emit twenty tons of carbon dioxide every year throughout their lives.[37]

This is a very odd approach. First off, the emissions expected in decades to come are estimated at an old, too-high emissions rate for the United States; it has already declined 20 percent. Moreover, official expectations are actually that per person emissions will drop a further 0.5 percent every year before 2050 (as research pays off and more of our daily activities become slightly more carbon efficient). But by far the biggest issue is that blaming a parent for emissions of *all* future generations is ludicrous. By this measure, our hunter-gatherer ancestors were far worse for the planet than any jet-setting billionaire today.[38]

Measured in reasonable terms, having a child means taking responsibility for putting into the world a new human who will produce about ninety years of carbon emissions. Whether your child, when she grows up, decides to have children of her own must be *her* choice. Thus, the total impact of having a child is at worst about 15 tons of carbon dioxide every year for her life, about ninety years. This is 1,350 tons. If you feel guilty about having a child, you *could* compensate for those carbon emissions by purchasing $8,100 worth of authorizations from RGGI.

The broader approach, telling people not to reproduce, is preposterous. For the vast majority of people throughout history, children have provided the meaning and happiness that make life worth living. Children cost us in all sorts of ways: missed sleep, emotional havoc, college tuition, and, yes, carbon emissions. And yet people keep having them, because the benefits are so much greater than the costs.[39]

THE NEXT TIME you read about the actions you "should" take to help the planet, consider if it is just another case of "if everyone does a little, we'll achieve only a little." The truth is that most of our personal actions can have only a tiny impact.

Let me be clear: I'm not saying that you shouldn't think carefully about your own personal choices. There are solid reasons why any of us might choose to change our diets, drive a smaller car, and reduce the carbon footprint we leave on the planet. But climate change shouldn't be the major consideration, because the effect of such choices is so limited.

Much as we may wish otherwise, personal actions are not going to fix climate change. So, let us turn next to the actions made by governments that actually *could* make a big difference: carbon-cutting policies and international agreements. Why aren't those working either?

WHY THE GREEN REVOLUTION ISN'T HERE YET

WAY BACK IN 1976, a leading environmental campaigner, who remains a loud voice on climate today, confidently declared that an economy based wholly on solar energy was "now economic or nearly economic." He was wrong then, and four decades later, he is still wrong. Governments around the world spend more than $140 billion every year subsidizing inefficient solar energy and wind power. Yet despite this huge expenditure, together these renewable sources provide only about one percent of global energy needs.[1]

So why hasn't the green energy revolution happened yet? Because without breakthrough innovations, it remains enormously expensive.

Carbon dioxide emissions are a by-product of the cheap and dependable energy delivered by fossil fuels that has underpinned two hundred years of development. Ending our reliance on fossil fuels within decades will cost hundreds of trillions of dollars. Most rich countries attempting to meet that price tag would face electoral upheaval. Instead, they settle for spending hundreds of billions of dollars subsidizing highly visible solar and wind, without achieving much. The world's poor nations don't have trillions to spend. For them, the prospect of getting *more* energy from fossil fuels is much more enticing.[2]

And thus, everyone on the planet talks about renewable energy, but little of any substance is happening.

———————

WE ARE CONSTANTLY being told that renewables like solar and wind are just about to take over the world. This is almost entirely wishful thinking. One of the foremost climate campaigners, Jim Hansen, puts it best: "Suggesting that renewables will let us phase rapidly off fossil fuels in the United States, China, India, or the world as a whole is almost the equivalent of believing in the Easter Bunny and Tooth Fairy." Bear in mind that Dr. Hansen is the climate scientist who initiated public concern on global warming when he testified to Congress back in 1988, and he was former vice president Al Gore's climate advisor.[3]

When most people talk about renewables, they typically think of solar and wind energy. These are the "new" renewables. But globally, 85 percent of all renewable energy comes from wood and hydropower, what we can call "old" renewables. These old sources have the benefit of providing power *when we need it*. In contrast, solar and wind power can't be turned on when needed. Despite the incessant hype, it remains an inconvenient truth that they produce power only when the sun decides to shine or the wind decides to blow. That is why solar and wind power work only as a small addition to the baseload power that comes from fossil fuels and other reliable energy sources.[4]

If we want to increase the use of solar and wind significantly, we need to add backup power, such as idle fossil-fuel-powered gas turbines (that can be turned on when the sun isn't shining or the wind isn't blowing) or batteries (that can store their energy). Both add significantly to the cost of solar and wind energy.

The sheer scale of battery storage capacity that would be needed for solar power to work is vastly underappreciated: today the United States has enough batteries across the entire nation to store just fourteen seconds of average US electricity use.[5]

These fundamental economic and technological challenges are why no big nation in the world is anywhere close to seeing new renewable energy do more than nibble at the edges of energy consumption. This is clear when you look at the share of renewable electricity from all sources in the US.

Wind power produced less than 7 percent of US electricity in 2018, and solar power less than 2 percent. In total, renewable energy sources produced 17 percent of the United States' electricity, but most of this

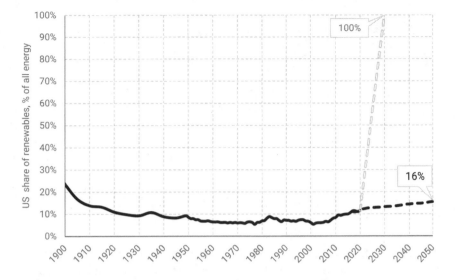

FIGURE 7.1 Percentage of US energy consumption (not just electricity) that comes from renewables, 1900–2050. The almost vertical line showing 100 percent renewables by 2030 represents common claims to achieve rapid conversion to all renewables.[6]

came from old renewables, with the majority coming from reliable hydroelectricity.

And electricity is itself only a part of all the energy used in the United States, which also includes the energy utilized in industrial production, in heating buildings, and in driving cars, trucks, and buses. When measured against all energy use, wind power in the United States produced 2.5 percent of US energy in 2018, and solar power only about half a percent. Almost three-quarters of renewable energy still came from old wood and hydropower.

As we can see in figure 7.1, over the past century the share of renewable energy in the United States declined from almost a quarter of all energy use to little more than 5 percent. Since 2000, mostly because of concern about climate change, renewable energy use has picked up from high single digits to about 11 percent today, and is expected to reach almost 16 percent of all US energy by midcentury according to the government's latest official estimate (almost unchanged from the last Obama administration estimate in 2017 of 16.5 percent).[7]

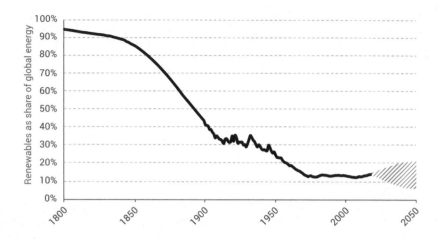

FIGURE 7.2 Renewable energy as a percentage of total global energy, 1800–2050. Data from 1800 to 2018. The hatched funnel shows the spread of scenarios from the International Energy Agency and United Nations.[8]

We need to be honest: reaching 16 percent of total US energy in 2050 does not mean that renewable energy is taking over from fossil fuels. That percentage is actually *less* than what US renewable energy sources contributed in 1900. And even by 2050, solar and wind power are projected to make up less than half of the 16 percent of renewables (four and three percentage points respectively), with most of this still coming from the reliable old sources of wood and hydropower.

When we hear politicians vowing to deliver "100 percent renewable energy" by 2030 or 2050, look at figure 7.1. We could call it the "narwhale chart"—it shows that such ideas are unmoored by historical reality or common sense.

The US experience reflects global trends. Today, solar panels and wind turbines together deliver only 1.1 percent of global energy. The International Energy Agency estimates that even by 2040, solar and wind power will cover less than 5 percent of global energy needs.[9]

The jarring fact is that humanity just finished spending two centuries *getting rid of* renewable energy and replacing it with fossil fuels (see figure 7.2). When everyone was poor, the whole world cooked and kept warm using polluting renewable energy sources like wood and dung. Over a century and a half, we shed our reliance on renewable energy and powered the industrial revolution with fossil fuels. For the last fifty years, the level of renewables globally has hardly budged from a level of around 13–14 percent. This reality mostly reflects the continued reliance of the world's worst-off on wood and dung.

Indeed, when we look at pathways ahead to 2040 or 2050, we see that it is very unlikely that the world will reach a point where more than 20 percent of its energy needs are met by renewable sources. It is likely that our share of renewable energy in 2050 will still be lower than what it was in 1950. It utterly beggars belief to suggest that we'll manage to get to 100 percent by 2030 or 2050.

IN THE POOR WORLD, replacing fossil fuels with new renewable energy sources like wind and solar power is hard because most people desperately want *much more* power at lower cost, not fickle power at high cost.

Half a century ago, the average Chinese had *less* electricity available than the poorest African today. Since then, with rapid industrialization and a rocketing use of coal power, China has catapulted itself away from poverty and increased its GDP per person seventy-two times, but at the cost of becoming the world's biggest carbon dioxide emitter.[10]

Most poor countries want to follow China's path. As Gwede Mantashe, South Africa's energy minister, puts it: "Energy is the catalyst for growth." Indeed, the International Energy Agency estimates that by using more coal, oil, and gas, and less renewable sources—along with much better governance—Africa can usher in an "African Century" that will increase carbon dioxide emissions but also grow the economy much faster, such that by 2040 every African will make $1,000 more every year.[11]

But why can't we solve the poor world's energy poverty problem with solar panels and wind turbines? This is in fact what many development organizations and green energy companies claim to be doing. Their claims do not add up.

To see how the benefits of green energy sources like solar and wind power are being massively oversold, we should look to the poor Indian village of Dharnai, which became India's first solar-powered community. The citizens had for years unsuccessfully tried to get connected to the national electric power grid, which mostly is supplied by coal-fired power plants. Along came a green benefactor. In 2014 under the slogan "Energy access simplified," Greenpeace supplied Dharnai with a solar-powered "micro-grid"—an electric grid that's not connected to the central grid. Greenpeace proudly declared that "Dharnai refused to give into the trap of the fossil fuel industry." The world's media reported excitedly on the "first village in India where all aspects of life are powered by solar."[12]

Greenpeace had the best of intentions. But intentions don't count for much in development. The day the electricity was turned on, the batteries were drained of power within a few hours. A boy from Dharnai remembers wanting to do his homework early in the morning before leaving to work in the fields, but there wasn't enough power for the family's one lamp. Because solar power is way too weak to power a stove, the citizens now had maybe a lamp illuminating the kitchen, but they still had to use the same old cookstoves burning wood or dung, polluting the home and putting the whole family at risk.[13]

Greenpeace invited the state chief minister to the inauguration of the solar power system so he could meet grateful inhabitants. When he showed up, he was met by a large crowd of people waving signs demanding "real electricity" (the kind you can use to run a refrigerator or a stove, and that your children can use to do their homework in the morning) and not "fake electricity" (meaning solar energy that could do none of these things). A week after the protests, Dharnai was hooked up to the national grid with more reliable electricity access. Its residents now get power from the grid at one-third of the price.[14]

This situation repeats itself around the world.

In Fiji, the government teamed up with a Japanese technology company to deliver off-grid solar power to remote communities. They provided a centralized solar power unit to the village of Rukua. Prime Minister Frank Bainimarama proudly declared he had "no doubt that a number of development opportunities will be unlocked" by the provision of "a reliable source of energy."[15]

Understandably, all of Rukua was thrilled to get access to energy and wanted to take full advantage. So more than thirty households purchased refrigerators. Unfortunately, the off-grid solar energy system was incapable of powering more than three fridges at a time, so every night the power would be completely drained. That led to six households buying diesel generators. According to researchers who studied this project: "Rukua is now using about three times the amount of fossil fuel for electricity that was used prior to installation of the renewable energy system." In rather understated language, the researchers conclude that the project did not "meet the resilience building needs" of the community.[16]

Solar panels do deliver some benefits, allowing you to charge your cell phone and run a light at night. But they don't deliver benefits that help drive development. One common anecdote holds that solar lights allow students to study at night and hence improve learning. Published in 2017, the first controlled study from India shows that solar power actually has no impact on time spent on studying or schooling. It also shows that solar power doesn't increase time spent working or improve savings, spending, business creation, or broader development. As a result, solar panels are not what people most want. In Tanzania, a survey of households with solar panels showed that almost 90 percent still wanted access to the electric grid.[17]

Moreover, when asked in polls across the continent, Africans report that solar panels provide much less benefit than they cost. On average, the private benefit is worth only 30 to 41 percent of the total costs. Even when including the extra health benefits like a reduction in damaging indoor air pollution from kerosene lamps, it is likely that the solar panels are worth less to the average African than what the well-meaning rich people pay to put them up.[18]

IN CLIMATE-EXCITED RICH COUNTRIES, governments have attempted to force more renewable energy into the system through regulation and subsidies. In Germany and the European Union as a whole, we see the results of this approach.

Germany's renewable energy policy, called the Energiewende, has been hailed by environmentalists and politicians around the world. Under this decade-old policy, the nation has moved away from nuclear and fossil fuels toward wind, solar, and biomass energy. The Energiewende has cost $36 billion annually in recent years, and is the country's biggest political project since reunification. Electricity costs have doubled over the past two decades and are now at 35¢ per kWh, almost three times the US average. Germans will have spent $580 billion on renewables and related infrastructure by 2025.[19]

This massive expenditure has meant renewable energy sources have gone from meeting 7 percent of Germany's total electricity needs in 2000 to 35 percent of electricity needs in 2019, with solar and wind power accounting for two-thirds of the renewable electricity. But the country has been worried about nuclear power, especially since a 2011 accident in Fukushima, Japan, triggered by an earthquake and tsunami. This fear has led to a halving of nuclear energy from 2000 to 2019. Since nuclear power is carbon free, the decline in nuclear energy has offset much of the increase in solar and wind power.[20]

Overall, fossil fuels have declined only slightly as a share of German energy. In the first decade of the new millennium, the fossil fuel share of the overall energy supply dropped somewhat from 84 to 80 percent. But in the years since the Energiewende was passed in 2010, the fossil fuel share has stayed almost constant, inching down just one percentage point to 79 percent today.[21]

In the larger European Union, renewable energy sources have since the turn of the century increased from 6 to 14 percent in 2018. But there's a catch. Most of this renewable energy does not come from solar and wind. In total, solar and wind make up 2.7 percent of all energy, whereas biomass makes up more than 10 percent. Biomass, which basically is a fancy-sounding name for wood, is one of the old, reliable renewables that can produce energy when it is needed. The problem for the planet is that wood is often imported from US forests in diesel-driven ships, and emits *more* carbon dioxide than even coal when it is burned. Biomass is categorized by the EU only as carbon dioxide free because it is hoped that felled trees will be replanted and over many future decades will soak up as much carbon dioxide as was released by its burning. Needless to say, this is dubious accounting at best.[22]

With this trickery and reliance on wood imported from the United States, the EU has managed to cut fossil fuel use from 79 percent in 2000 to 71 percent in 2018. Yet, the cost of this energy policy is now more than two percent of the EU's economy, or about $400 billion every year.

Indeed, about 20 percent of the EU budget is now being spent on climate policies.[23]

Today, residential electricity costs in the EU are twice the US cost of 13¢ per kWh. This gap will grow quickly. Over the next decade, one 2019 study estimates that mostly because of stronger climate policies, the wholesale electricity price in the EU will *quadruple*.[24]

IT IS IMPOSSIBLE to get around the fact that climate policies are expensive. What do campaigners and politicians do? They either downplay the cost, or far more dangerously, maneuver to make it appear that there will actually be a net benefit.

If we change from dirty fossil fuels to clean renewable sources, they claim, we will not only solve the climate crisis but unlock jobs, savings, competitiveness, and improved well-being. American political commentator Thomas Friedman frequently talks about how the United States needs to jump on the clean energy bandwagon or face the prospect that China will "clean our clocks" when they take home all the "benefits."[25]

This attempt to rewrite the facts is perhaps nowhere clearer than when in 2018 the UN secretary-general touted a report that he said showed the world could gain "at least $26 trillion" in benefits from strong climate action as soon as 2030. What are we waiting for?[26]

Weirdly though, the report claimed $26 trillion in benefits, without explaining how that number was reached. The *actual* documentation, the report told us, was in a working paper that would be published shortly. I contacted the authors and asked to see how they reached this fantastical figure. For a year and a half, I have been in touch every few weeks with people connected with the publication. At the time of writing this book, the document has yet to be made available.

Claiming that climate policy is not only good for climate but also will actually make everyone rich is a comforting bedside story. But it is flat-out wrong. Every serious report shows extraordinarily large costs from climate policy, simply because changing the energy infrastructure that has underpinned the last two centuries of economic growth will be very, very costly.

The UN estimates that the additional infrastructure cost alone of achieving the 2.7°F (1.5°C) limit would come to almost a trillion dollars each and every year for the next thirty years. A 2018 Goldman Sachs report shows that the cost to build just the infrastructure for electric cars, such as charging stations and power networks, would be an astronomical $6 trillion, or 8 percent, of today's global GDP.[27]

As such costs rack up, political opposition will become ever more likely, causing backtracks on climate policy as we have seen in France and elsewhere, and even cancellation as President Trump has done with US involvement in the Paris Agreement.

If the European Union sticks to its climate promises for 2050, it alone could end up paying more than $2.5 trillion per year in climate costs—10 percent of its entire GDP. This is more than all the EU's current spending on education, health, environment, housing, defense, police, and courts. It is inconceivable that such spending will go unchallenged.[28]

NEW RENEWABLE ENERGY sources like solar and wind cost $141 billion annually in subsidies globally, and matter little in the global energy supply. Rich countries can spend trillions and still achieve very little. Poor countries don't have trillions to spend, and instead want more energy, which will come predominantly from fossil fuels.

This, in a nutshell, is why global climate policies are failing. There has been far more global attention to climate change in recent years, with protest movements and campaigners ensuring the issue remains on the front page of newspapers. But the world's decision makers are even more off track than ever. This is why the UN itself summarized the 2010s as a "lost" decade, and found there was essentially no difference between contemporary reality and an imaginary scenario in which there had been no new climate policies since 2005. This was a decade full of global talks, of promises made by politicians, and of climate laws enacted. Yet, on a global level, the UN could not see any of this action making any difference whatsoever.[29]

Next, we will explore exactly why the Paris Agreement is not going to provide the salvation that was promised.

WHY THE PARIS AGREEMENT IS FAILING

IN DECEMBER 2015, leaders from almost every country in the world approved the Paris Agreement on climate change. It was hailed as a landmark achievement, tackling climate change through national promises of carbon dioxide cuts. The agreement's preamble even spoke of limiting temperature rises to less than 2°C (3.6°F), or even 1.5°C (2.7°F). French president François Hollande said: "This is a major leap for mankind." Economist Lord Stern added: "This is a historic moment, not just for us but for our children, our grandchildren and future generations." Al Gore saw it as "bold and historic."[1]

Unfortunately, they are wrong. The Paris Agreement will cost a fortune to carry out and do almost no good.

ALTHOUGH ITS NAME suggests there is just one elaborately engineered "agreement," in fact, the deal is made up of national commitments from individual countries that vary wildly; each country simply declared how much carbon dioxide emissions it would reduce until the year 2030. Some countries made ambitious promises, others made far easier to achieve vows. The Paris Agreement itself effectively staples together all those promises.

Each of those promises has a cost. To achieve the promised carbon cuts, policies will have to make people and companies use less carbon-dioxide-emitting technologies and fuels. If these technologies and fuels were cheaper, obviously this would be cost free, but the cuts would also happen by themselves and there would be no need for any promises. The fact that nations are making these promises and that they are difficult to achieve means the policies will have to force people and companies to use more expensive technologies and fuels.

The governments themselves rarely add up the cost of these promises, but it is possible to work out. Unfortunately, it is not as easy as just looking at energy taxes or subsidies. If a nation's policy means its citizens have to pay, say, $10 billion in extra energy taxes, then it may seem like the cost is simply $10 billion. But that $10 billion doesn't disappear; it is collected by the state and used to fund other things in the budget. Equally, when a government spends $10 billion in subsidies on solar and wind power, that doesn't mean a societal cost of $10 billion, because the money is not wasted; it is redistributed, often to the rich owners of the solar panels and wind turbines.

So what is the real cost? We need to identify the knock-on effects of higher energy prices for everyone. Every household, business, and organization that uses energy finds it a bit more expensive and has a little less money for other things. This slightly slows economic growth. This cost is the relevant social cost of climate policies—the reduction in welfare that comes from each nation insisting on using energy that is slightly more costly and less reliable than fossil fuels.

Emissions of carbon dioxide are largely by-products of productivity—of industry, governments, and individuals producing things that we want more of (including heating, cooling, food, transport, hospital care, and so much more). Wishing this wasn't so doesn't make it go away. When countries promise to reduce their emissions, they are effectively promising to make all these things a touch more expensive. That acts as a slight brake on the economy, leading to a small reduction in growth. It does *not* mean that countries won't grow; it simply means they will grow slightly more slowly.

WE CAN MEASURE the cost of this slowing growth using energy-economic models. Put simply, each model can identify the likely pathway of GDP growth over coming decades both with and without the climate policy. The difference in GDP between the two scenarios is the cost of the climate policy.

Now, any model is only as good as the data and assumptions put into it. One model might be overly pessimistic. It might be based on an assumption that it is very hard to cut emissions, so strong climate policies will result in substantially smaller GDP increases, whereas another might be overly optimistic, based on an assumption that all emission cuts can be achieved at almost no cost. Unsurprisingly, climate policy partisans reliably pick the model that advances their arguments. That is why economists prefer to use multiple models, to average out the optimists and pessimists.

The most prominent, nonpartisan research program that looks at the findings from a large number of energy-economic models is the Energy Modeling Forum (EMF) based at Stanford University. The EMF has conducted more than thirty such studies and is considered the gold standard of energy-economic modeling. Using multiple studies, mostly from the EMF, we can estimate the cost of the most expensive promises made under the Paris Agreement: those made by the United States, the European Union, China, and Mexico. Together, these commitments make up about 80 percent of the total promised carbon reductions.

In promises made under President Obama, the United States said that by 2025 it would reduce its overall greenhouse gas emissions by 26 to 28 percent compared to the emissions in 2005. Work from the EMF shows that effective policies to achieve this goal would have increasing costs over the period. In 2015, when the promise was made, obviously there would be no cost. But as the policies to achieve the promise were phased in, GDP would grow slightly slower than it otherwise would have. The increasing gap between the GDP that is and the GDP that would have been is the annual cost. By 2030, it is estimated the promised cuts would result in a GDP loss for the United States between $154 billion and $172 billion.* As the US GDP is now on a slightly lower growth path, this loss would continue into the future.[2]

---

*Under the Trump presidency, the United States has announced a decision to leave the Paris Agreement. However, this will only take effect in late 2020 after the presidential election, so at the time of this writing, the US is still a signatory bound by its promises.

The EU promised to cut its emissions by 2030 by 40 percent compared to its emissions in 1990. There is no official estimate of the cost, but the EMF finds across seven models that reducing emissions by 40 percent in 2030 (as a pit stop toward an 80 percent reduction in 2050) leads to a GDP loss of 1.6 percent in 2030, or $322 billion.[3]

Not every country took the same approach to making promises. China vowed to reduce the amount of carbon dioxide emitted for each dollar produced across its economy. That's known as "carbon intensity." It set a 2030 target of reducing the carbon dioxide emitted per dollar generated by its economy by at least 60 percent compared to 2005. This will be equivalent to reducing its overall emissions in 2030 by at least 1.9 billion tons of carbon dioxide, or 1.9 gigatons (Gt) in short.* Here we can use the results from the Asia Modeling Exercise, a research project similar to the EMF but focused on the region, which ran thirteen different energy-economic models with and without climate policies. Its results suggest that China can reduce 1.9 Gt of carbon dioxide for about $200 billion in annual GDP loss.[4]

Mexico has enacted the strongest climate legislation of any developing country; it has conditionally promised to reduce its emissions by 40 percent below what it would otherwise have emitted by 2030. Although Mexico itself has low-balled its cost estimates, independent researchers using several models predict the cost to reach 4.5 percent of GDP by 2030, or about $80 billion annually.[5]

This means that the total cost to the United States, the EU, China, and Mexico adds up to $739 billion (or $757 billion if the US goes for the higher end of its promised range). Given that these countries are responsible for 80 percent of the promises in the Paris Agreement, it is reasonable to assume that the $739 billion constitutes 80 percent of the pact's total cost. That makes the annual global cost of the Paris Agreement by 2030 about $924 billion.

---

*Since carbon dioxide is the most prominent greenhouse gas, we often just talk about carbon dioxide, but there are many other greenhouse gases, like methane. To keep life simple, we are converting all of those gases according to the international standard into carbon dioxide, or what is technically called "carbon dioxide equivalents."

Remember, all of these cost estimates assume that politicians will implement the most effective policy to achieve their nation's promised emissions reduction; typically, *one* carbon dioxide tax applied across all sectors of the economy, slowly increasing over time. In real life, that never happens. Policy makers love picking winners and making special deals.

For instance, back in 2008, the EU promised to cut its emissions by 20 percent by 2020. Stanford's Energy Modeling Forum ran a number of models for this policy and found that if done effectively, the EU would end up with a GDP in 2020 about 1 percent lower than otherwise: the cost of the policy would be 1 percent of GDP forever from 2020. "Effectively" in this case would have meant using one carbon dioxide tax that would mostly have encouraged power producers to switch from coal to less polluting natural gas. Unfortunately, politicians couldn't help but push their economies to try to increase the amount of wind and solar power utilized. They used subsidies that made the policy's cost much more expensive. In total, the researchers have found that the total EU policy will end up costing around 2.2 percent of GDP, more than double what the price tag could have been had the EU's politicians made more effective choices.[6]

We see the same problem in the United States, where many politicians not only want power grids to emit less carbon dioxide, but also want the reduction to be delivered by renewables: fundamentally, solar and wind power. In twenty-nine states this is achieved through a so-called renewable portfolio standard that obligates power producers to get a specified fraction of their electricity from renewable sources. California and New York have both set 50 percent targets for 2030, while Hawaii aims for 100 percent by 2045. While this type of policy is obviously popular with solar and wind producers, research shows that it ends up costing states extra. Whereas the cheapest emission reduction, just like in Europe, will often be a switch from coal to gas, relying on renewables at least doubles the cost for states.[7]

The Paris Agreement is unlikely to be an exception to the rule that politicians do things less efficiently and less effectively than they could. The promises made by various nations will be implemented ineffectively, just like those made by the EU and the United States. Therefore, it is plausible that the Paris Agreement costs will be twice what they would be under the most effective policy possible.

Thus, without hyperbole we can say that the Paris Agreement will easily set the world back by at least $1 trillion annually by 2030—and more plausibly, with less efficient policies, the cost could climb to somewhere close to $2 trillion annually.[8]

No matter which way you look at it, the Paris Agreement is by far the most expensive pact in history. At $2 trillion, it is at par with the entire expenditure on the world's military each year. Every year, the Paris Agreement will cost about two to five times the total amount of the world's previously most expensive global accord: Germany's World War I repayment settled in the Versailles Treaty. And to put this in context, compared to some of the other policies undertaken to help the world, the annual cost of Paris is about a hundred times the amount the world annually spends on protecting and promoting biodiversity. It is also about a hundred times higher than the amount the world annually spends on all policies to tackle HIV/AIDS.[9]

BY THIS POINT, you'll probably not be surprised to learn that there is no official estimate of what the Paris Agreement will actually *achieve*.

To evaluate the impact on climate change of the promises made under the Paris Agreement, first we need to identify the base level of emissions today. Then we can work out what the promises of carbon cuts mean, in terms of effect on temperature rises.

From 2020 to the end of the century, if we do nothing, the standard UN business-as-usual scenario sees the world emitting about six thousand gigatons of carbon dioxide.

The total number of tons we emit is directly connected to average, global temperature rises. The UN estimates that an extra 1,000 Gt of carbon dioxide emitted will result in a 0.8°F increase over the long run. That means that an extra 6,000 Gt over the century will result in a temperature rise of about 5.7°F or so. Since we're already almost 2°F above preindustrial levels, that means that the twenty-first century will see the planet get roughly 7.5°F hotter if we do nothing, similar to what we saw with the MAGICC model in chapter 2.[10]

What will happen, then, if nations meet their promises under Paris? The United Nations organizers of the Paris Agreement once in 2015 (and never since) released an estimate of the total maximum impact of all carbon dioxide cuts promised by all nations. It provides the absolutely best-case scenario that we can hope for. This estimates a total reduction of 64 Gt carbon dioxide through to 2030. According to the UN's estimate of 0.8°F per 1,000 Gt carbon dioxide, this translates to a reduction in temperature by the end of the century of about 0.05°F. What this tells us is that even in an optimistic scenario, the Paris Agreement isn't going to come anywhere close to solving global warming. It will have a miniscule impact on the temperature by 2100.[11]

IN FIGURE 8.1, we see the feeble results of the Paris Agreement, compared to what would be needed to limit temperature rises to 2.7°F. The agreement promises only a tiny reduction of sixty-four gigatons of carbon dioxide, or a 0.05°F temperature reduction. And this is only in a best-case scenario where every nation achieves everything that it has promised. But in reality, we are not on track to achieve even that.

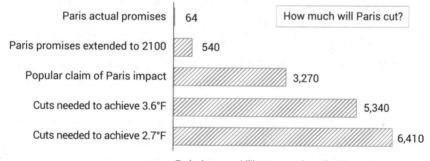

Emissions cut, billion tons carbon dioxide equivalents

FIGURE 8.1 The size of different emission cuts. The first bar shows how much the Paris Agreement will cut by 2030. The second bar shows what the agreement will cut if promises are kept throughout the entire century. The third bar shows how much its popular claims suggest it will achieve: fifty times more than the *actual* Paris Agreement. For comparison, we see the cuts needed to limit temperature rises to 3.6°F (2°C) and 2.7°F (1.5°C). All these numbers have large uncertainties and should be used for comparison of relative size.[12]

A 2017 landmark article in *Nature* puts it bluntly: "All major industrialized countries are failing to meet the pledges they made to cut greenhouse-gas emissions." The researchers noted that the European Union promised to cut to 40 percent below its 1990 level by 2030, but had enacted policies that would reduce less than half that, to 19 percent. Even including pledged policies, the EU will make it to less than 30 percent by 2030. And what progress it makes is mostly coming from a shift from coal to gas, rather than from the construction of renewable energy.[13]

The story is the same around the globe: although it promised to cut emissions by 18 percent below 1990 levels by 2030, Japan is on target to cut by just 4 percent. Emissions in Mexico and South Korea have hardly shifted, despite their promises. And it would be entirely wrong to imagine that the United States was on track before President Trump announced plans to quit the Paris Agreement. President Obama promised in Paris to cut emissions by 18 percent below 1990 levels by 2025, but never backed this with sufficient legislation. Based on Obama-era policies (and not taking account of Trump's reversal of them), President Obama was on track to achieve, at most, a 7 percent reduction.[14]

In contrast, the world's developing countries are on target, but only because they promised so little. For example, India pledged to reduce emissions by so little that its promises are likely to be achieved without any new climate policies at all, and it will have met its obligations even if its emissions increase by more than fourfold from 1990 to 2030.[15]

A 2018 study finds that of the 157 countries to have promised emissions cuts in the Paris Agreement, only seventeen have passed laws to do so. In other words, only one in ten countries is taking the necessary steps to get on track to achieve what they promised. These are not the biggest countries. The nations are Algeria, Canada, Costa Rica, Ethiopia, Guatemala, Indonesia, Japan, Malaysia, Mexico, Montenegro, North Macedonia, Norway, Papua New Guinea, Peru, Samoa, Singapore, and Tonga. The world will not be saved by emissions cuts in North Macedonia, population two million, and certainly not by the most drastic actions that could be taken by Tonga, population 108,000.[16]

Every single major industrialized country is failing to live up to the promises it made under the Paris Agreement, and the few countries on track are too small to make any significant impact at all.

SUPPORTERS OF THE Paris Agreement sometimes accept that the promises through to 2030 will achieve next to nothing, but then argue that further down the road, the pact has the potential to achieve much, much more. To justify this boosterism, they imagine that countries will change course and actually start delivering on their pledges, and that they will go even further after 2030.

So, what if we give the politicians the benefit of the doubt? What if we imagine that governments break with historic precedent and honor their commitments, even though they failed to honor commitments made in past climate deals? What if we even assume that they will continue to honor the Paris Agreement pledges not just until 2030, but *for the rest of the century?*

This is incredibly unrealistic. The two previous global climate agreements in Rio de Janeiro in 1992 and in Kyoto in 1997 came to almost nothing at the end of their deal periods, and certainly they were not prolonged for another seventy years. Nonetheless, it's worth setting out such an artificial scenario just to illustrate what we would achieve, in this hypothetically "best case" scenario.

Extending the Paris commitments to 2100 would deliver cuts equal to 540 Gt of carbon dioxide. Translated into an effect on temperature, it would reduce the increase in temperature rises by the end of the century by a tiny 0.4°F.[17]

To justify their support for the Paris Agreement, some backers assume not only that all of today's promises will be delivered, but that future climate conferences will lead to global pacts with ever larger carbon-cutting promises. They assume all of *those* promises will also be delivered. And then they attribute *all* of those imaginary, future reductions to the Paris Agreement. The global warming policy campaigners who run the influential Climate Action Tracker website make all these assumptions, and as a result they conjure fifty times more reductions from the Paris Agreement than were actually promised in Paris.[18]

This approach is astonishingly misleading. It's as if I promised to skip eating a piece of cake today and told you that as a result I would lose twenty-five pounds over the next year. No, that result requires my not only making a promise for one day, but also taking action over twelve entire months. *Promising* to take a first step is important but it isn't the same thing as starting, and it certainly isn't the same as finishing the entire journey.

At the Paris climate summit, politicians also promised to keep global temperature from ever rising beyond 3.6°F (the so-called 2°C limit). Accomplishing this would mean cutting eighty times more carbon emissions than were actually promised (and remember, they are not on track to achieve even that much). To achieve the 3.6°F target, we would literally need new and additional carbon cuts of the same size as those in the Paris Agreement *every single year* from 2020 to 2100 (and we would need them to be actually delivered, not just promised). The politicians even went so far as to declare in their agreement that they wanted to limit the global temperature increase to below 2.7°F (1.5°C), which would mean cutting one hundred times more than what was promised in Paris. It is therefore accurate to say that the Paris Agreement—even if actually achieved—provides us with only one percent of what politicians are promising.

Currently we have promised to spend $1–$2 trillion every year, and we won't be able to tell the difference in temperature even in a hundred years. Indeed, it turns out that if we measure all the benefits of reduced climate damage in monetary terms, every dollar the Paris Agreement costs will avoid just 11¢ worth of long-term climate damage. That isn't sensible.[19]

SO OFTEN TODAY we hear campaigners and politicians say the real problem is simply that promises have not gone far enough. Indeed, since the signing of the Paris Agreement it has become more and more fashionable to suggest that entire nations should go net-zero; that is, stop contributing carbon dioxide to the atmosphere by 2050 or sooner. More than sixty countries have promised to achieve carbon neutrality within the next thirty years. The biggest emitters—China, the United States, and India—are not among them. The biggest nations that have made the pledge are the UK, France, and Germany. Finland has said it will aim for carbon neutrality by 2035, while Norway is aiming for neutrality by 2030. Australia hasn't made a net-zero commitment, but four of its states have.

Within the US, New York and California are aiming to achieve carbon neutrality by 2050.[20]

All this is far easier said than done. It is going to be very expensive. And, in every case, the promise is likely to be broken.

It is instructive to look at the case of New Zealand. It was actually the first country in the world to promise to go carbon neutral. It is also the first country to have spectacularly failed, and the first to promise for a second time to achieve the same thing.

In 2007, Prime Minister Helen Clark declared her vision was that the small nation would become carbon neutral by 2020. She was celebrated by the United Nations as a "Champion of the Earth." If only cutting carbon was as simple as winning attention. New Zealand not only failed to achieve the vision, but also failed even to reduce *any* emissions. The latest 2019 official statistics show that the country's total emissions will be *higher* in 2020 than they were when Ms. Clark's ambition was declared. New Zealand is on track to be a whopping 123 percent off Ms. Clark's vision. Yet, in 2018, Prime Minister Jacinda Ardern reupped the pledge, promising to achieve carbon neutrality by 2050. Legislation aimed at achieving that goal was passed in 2019.[21]

New Zealand is a fascinating case study because, to its credit, Ardern's government actually asked its leading economic authority to estimate the cost of her promise. Thus, we have what is likely the only official, academically credible estimate of what it will cost to achieve carbon neutrality. This research, undertaken by the leading independent economic think tank in New Zealand, shows that just getting halfway to the target—cutting 50 percent of New Zealand's emissions by 2050—would cost at least $19 billion annually by 2050. For a small country with a population similar to that of the Republic of Ireland or the state of South Carolina, that's a big deal, about what the government spends now on its entire education and health care system.[22]

And it is only the cheapest cost of getting halfway to Ardern's target. Getting all the way will likely amount to more than $61 billion annually, or 16 percent of GDP by 2050. That is more than New Zealand today spends on social security, welfare, health, education, police, courts, defense, environment, and every other part of government combined.[23]

To achieve their promise, New Zealanders will need to accept an escalating carbon tax that ends up so phenomenally high that it would be equivalent to a gasoline tax of $8.33 per gallon. And even the 16 percent GDP cost relies on a fairy-tale assumption that every single policy will be enacted as efficiently as possible. Bearing in mind the evidence that costs double in the real world, it could be 32 percent or more.[24]

The cost doesn't just start in 2050, which would make it easy to ignore. Getting there requires policies starting in 2020, meaning the costs will start coming in now, ramp up to 16–32 percent in 2050, and stay there for the rest of the century.

Across the century, the cost adds up to more than $5 trillion and could reach beyond $11 trillion. If we imagine each New Zealander paying an equal share of this amount every year across the century, the cost would be the equivalent of at least $12,800 for every single New Zealander, every year. If the policies are done badly, as they have been done so far across the globe, the cost per person could even go beyond $25,000 per year.[25]

As a back-of-the-envelope exercise, if we took the percentage cost of going carbon neutral in New Zealand by 2050 and applied it to the United States, that would imply a cost of at least $5 trillion in today's money. Not just once, but *every single year*. That is higher than the *entire* current federal spending of $4.5 trillion. And again, under realistic assumptions the amount could be closer to $10 trillion a year.[26]

But at least New Zealand will help the world in dealing with climate change, right? Even if the country will be going through a huge and protracted, self-inflicted cost, it will also deliver some good? Yes. But barely.

Let's get a sense of the size of the impact. If we assume that New Zealand this time will actually deliver on its net-zero promise in 2050 and stick to it throughout the rest of the century, the total amount of greenhouse gas reduction will, according to the standard estimate from the UN's climate panel, deliver a temperature reduction in the year 2100 of 0.004°F, or about four-thousandths of one degree Fahrenheit. Given the expected temperature increase by around 2100, this means that New Zealand going net-zero by 2050 will postpone the warming that we expected to see on January 1, 2100, by about three weeks to January 23, 2100.[27]

So, New Zealand is considering spending at least $5 trillion to deliver an impact by the end of the century that will be physically unmeasurable.

This will make it hard to get Kiwis to support such strong climate policies continuously for the next eighty years. Sooner or later, and likely sooner, a politician is successfully going to argue to dump the net-zero promise that will deliver zilch in a century, and instead double spending on things like health, education, and environment, *and* get some tax reductions.

THERE HAS NEVER been an official estimate of the cost of the Paris Agreement, nor has there been one that gives a meaningful evaluation of its impact. Looking at the numbers, it is obvious why.

The Paris Agreement will be the costliest pact ever agreed to, by far. It will cost us $1–$2 trillion per year from 2030 onward, if actually fully implemented. Yet the agreement will do almost nothing for the climate: all of its promises will reduce the temperature rise by the end of the century by an almost imperceptible 0.05°F. And none of the big emitting countries are anywhere close to actually delivering on their promises.

Spending trillions to achieve almost nothing is, not surprisingly, a bad idea. Every dollar spent will produce climate benefits worth just 11¢.

Surely, we can do better. Right now, we're pursuing a policy that won't solve climate change—not even close—and that will waste trillions of dollars along the way.

PICK A PATH:
WHICH FUTURE IS BEST?

IMAGINE IF EIGHTY YEARS AGO, in 1940, you had a choice. Looking ahead through wars, upheavals, discoveries, and opportunities, you could have set the United States on track to have either a slightly higher or a slightly lower average economic growth rate. It wouldn't be by much. Per person, it would mean a growth rate of either 1.27 percent or 1.89 percent—just 0.62 percentage points difference.

It would seem obvious to most people to pick the higher growth rate, making sure we would provide our children, grandchildren, and their children and grandchildren with more possibilities and opportunities. Yet, many people would probably be surprised by the magnitude of this small increase in growth over time. Because growth rates compound over the years, the higher growth rate would deliver a nation of individuals who would be two-thirds richer in just eight decades.

This is the choice the rich world faces today, when discussing the climate. And the choice is even starker for the world's poor. Spurred on by activists and campaigners, many global leaders are poised to pick the lower-growth pathway, condemning our children and grandchildren to a worse existence, and ensuring that the world's poorest are trapped in a future with fewer opportunities, less prospects, and less welfare, to the tune of $500 trillion. Per year.

THE GLOBAL APPROACH to cutting carbon dioxide has mostly failed for three decades. Yet, the climate policies resulting from that approach are likely to cost us a fortune. By 2030, the Paris Agreement will cost $1 trillion or more per year. But this is just the start. Climate panic is likely to end up costing humanity *hundreds* of trillions of dollars, every single year.[1]

This sounds like an extreme claim, but it is based on studies undertaken over a number of years by researchers working for the United Nations climate science panel, the IPCC, and published in 2017. They examined five different plausible scenarios for the future, called the Shared Socioeconomic Pathways, or SSPs. The pathways rely on complex modeling with a huge range of different inputs and assumptions to establish various broad, plausible futures. The researchers who created these models looked at factors ranging from geopolitics to economics, trade, migration patterns, education, and health.[2]

All of the pathways show a world in which people get richer over the coming century. That prediction is soundly based on history: we spent the last two hundred years growing the average global GDP per person by 1,600 percent, from $1,100 to about $17,500 today. But, depending on which pathway we choose, there are vast differences in *how much* wealthier people will be.[3]

The lowest increase takes place if we go down a pathway the researchers call Regional Rivalry—A Rocky Road. In this vision of the future (see figure 9.1), nationalism surges in many countries. That means there is less interest in a shared response to global warming. What's more, governments focus more on domestic challenges and national security instead of working together or investing in education and technological development. As a result, economic development is sluggish, and income inequalities persist or worsen over time. Population growth is low in industrialized countries and high in developing ones. The environment is given a low priority globally, meaning that in some regions environmental challenges worsen. Even in this depressing and fractious future, average GDP per person in 2100 will be 170 percent of today's level.

The second-worst outcome occurs if we go down a pathway called Inequality—A Road Divided, where the world splits into two: rich, well-educated countries on one side, and poor, less-educated countries with labor intensive, low-tech economies on the other side. While there

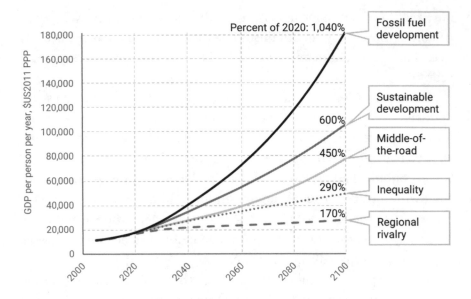

FIGURE 9.1 GDP per person across the world for the UN's five scenarios, 2005–2100. To the right, the GDP per person in 2100 in percentage of its 2020 value.[4]

are large investments in health and education in the rich world, poor countries are unable to make these investments, and increasing disparities in economic opportunity and political power lead to a growing gap between them. The gap between the rich and poor worsens within countries, too, so social cohesion degrades. Conflict and unrest become increasingly common. Rich countries make investments in the environment, but poor countries do not. Even in this grim scenario, by the end of the century average GDP per person will grow to 290 percent of today's level.

In the Middle of the Road scenario, social, economic, and technological trends look very similar to their historical pattern. It is a scenario not dissimilar to today, where many problems remain and many solutions are underfunded. There is slow progress in health and education, and technology advances without major breakthroughs. Development and income growth proceed fairly unevenly, with some countries falling behind. Global and national institutions continue to make slow progress in achieving the Sustainable Development Goals, the UN's global development agenda. Yet, as we bumble our way ahead, growth over the next eight decades will deliver GDP per person that is 450 percent of today's. In a 2018 survey of prominent economists, this outcome is actually seen as the most likely outcome for the twenty-first century.[5]

And then there are two other pathways: the two "best" scenarios, leading to the highest income growth. They offer us two completely different, conflicting approaches to making the world better off in 2100.

One is the Sustainable Development or Green Road pathway. This is an eco-scenario in which there is more global harmony and cooperation. The world focuses on reducing the environmental impact of consumer consumption and ensuring that economic growth creates a smaller footprint on the planet. Governments worldwide make big investments in education and health along with rapid technological development. Nonetheless, many policies are designed to ensure that people use fewer resources and less energy, such as by cutting meat diets and lowering consumption. As a result, total man-made greenhouse gas emissions are lower and the scenario sees the lowest temperature rise of all five pathways. This world will deliver a GDP per person that is 600 percent of today's level.

The final pathway is called Fossil-fueled Development or Conventional Development. Down this pathway, the world's sharp focus is on growth. This is achieved through strong, competitive markets and policies that foster innovation and build human capital through massive investment in health and education. The world focuses on rapid technological development. Countries exploit abundant fossil fuel resources to support resource- and energy-intensive lifestyles. Greenhouse gas emissions are much higher and this scenario sees the highest temperature rise by 2100. Nonetheless, much more is invested in adaptation, and local environmental problems like air pollution are successfully managed. Here, average GDP per person soars more than tenfold, hitting 1,040 percent of today's GDP per person.[6]

Both of these scenarios will see the average person on the planet in 2100 with GDP per capita over $100,000. That may sound fanciful to us today, but top economists give these final two pathways a one in three and one in four probability, respectively.[7]

Both of these scenarios will deliver amazing welfare for the average person, but the fossil fuel pathway delivers much more. Both of these scenarios will also see extreme poverty almost eradicated by midcentury, but the fossil fuel pathway will do so sooner—over the next three decades, the green scenario will see twenty-six million more extreme poor every year. And both of these pathways will see a dramatic decline in inequality between nations. All of the five pathways expect inequality to drop, as the poorer world partly catches up with the richer world. But especially the green and the fossil fuel pathways will see a dramatic reduction in inequality between nations. By 2100 under both scenarios, inequality will be lower than it has been for the past two centuries. And perhaps surprisingly, the fossil fuel scenario will be even more equal than the green scenario.[8]

The research identifying different pathways gives us a really clear way of looking at the options in front of us. These are all scenarios that the researchers deem plausible from where we are now. It's reassuring to note that none of the options is a *Mad Max*–type dystopian future. Even in the worst of the five scenarios, people end up better off than today. But we should be aiming to ensure that future generations are in the best situation possible. So with that in mind, we can dig deeper into the differences between the five pathways.

PERHAPS THE MOST important point that is clear from these scenarios is that when governments invest more in education, health, and technology, the world becomes tremendously better off, with less poverty and reduced inequality.

If nobody invests in education, health, and technology (as happens in "regional rivalry"), or only the world's well-off countries do (as in "inequality"), the world ends up worse off, more unequal, and with many more poor than in the other scenarios. If governments invest modestly in all three areas (as in the "middle of the road" scenario), there are modest improvements in income, inequality, and poverty. But significant investment in education, health, and technology (as in the "green" and "fossil fuel" futures) means that the world will be a much better place by 2100, with very high incomes, virtually no extreme poverty, and little inequality between nations.

In the worst scenarios, the global spread of education crumbles. In the bottom two scenarios, only about half of all adults will get a high school education or higher, and a quarter will be without any education by the end of the century. Illiteracy, which is at 12 percent today, will increase to almost one-fifth of the world. Globally, life expectancy will increase at most by a couple of years over the next eight decades.[9]

In the two best scenarios, almost everyone will have a high school or university degree, illiteracy will have been almost completely eradicated, and life expectancy will reach nearly one hundred years of age.

The most important policy choices in these scenarios have nothing to do with climate change. We need healthy populations that are highly educated, able to develop, and use advanced technology. If we can manage this vital challenge, it is likely that the end of the century will see a world somewhere between six and ten times richer, much more equal, and which has long eradicated extreme poverty.

IT IS NOW OBVIOUS that policy choices matter hugely in determining what the future looks like. The wrong decisions will make us poorer and less literate; the right ones will make us richer, more educated, and healthier. But which decisions will be best for us as citizens of planet Earth?

If we asked most pundits today which scenario they prefer—the sustainable, green scenario or the traditional, fossil-fuel-driven scenario—it seems fair to expect that the overwhelming majority would go with the green scenario in a heartbeat. We have become so focused on global warming that the choice seems blatantly obvious.

But let's drill down into the differences between the two, because there is a huge disparity in outcomes.

Going down the "green" pathway, the average person in the world will be six times as rich in 2100 compared to 2020 (see figure 9.2). Average income in 2100 will be an astounding $106,000 in today's currency. That is a great achievement.

Remember, being better off correlates to better health outcomes and greater life satisfaction. More prosperity means more overall welfare: a

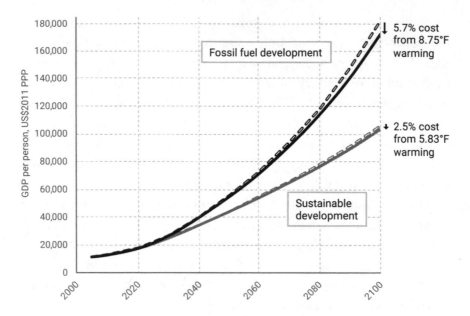

FIGURE 9.2 Global GDP per person, 2005–2100, for the green and the fossil fuel scenarios. The dashed lines show the GDP per person in dollars. But higher temperatures also mean more climate damages. The full lines show the GDP per person after deducting climate damages.[10]

general improvement in well-being and satisfaction. So we can reasonably say that overall global welfare is six times higher in 2100 under the "green" pathway. That's a great scenario.

But down the "fossil fuel" pathway, the average person will be 10.4 times as rich. That is a lot better. On average each person would earn $182,000 per year. This is $76,000 *more* per person per year in 2100. This is much better than the "green" scenario. A ten times increase in welfare is an astonishing success. Of course, it's very reasonable to worry that the "fossil fuel" pathway will actually increase human misery once we take global warming into account.

Using the damage estimates that William Nordhaus found, based on the work by the United Nations as we discussed in chapter 5, we can identify the climate impact on global welfare. The scenarios tell us how much warmer it will be: the green pathway takes us to a moderate temperature rise of 5.8°F by the end of the century, and the "fossil fuel" pathway leads us to a much warmer rise of 8.7°F. These levels of temperature rise will mean that the actual prosperity in the "green" scenario will be slightly lower (2.5 percent) by the end of the century, as seen in figure 9.2. Put into financial terms, the climate damage per person per year will be about $3,000. Yet, the remaining prosperity or welfare per person in 2100 will still be worth a fantastic $103,000 per person.

The temperature rise will be greater in the fossil fuel scenario and hence the damage will be bigger too (at 5.7 percent) at the end of the century, adding up to about $11,000 per person per year. Deducting this cost from the expected GDP per person means a welfare of $172,000 per person per year in 2100.

This is important and bears repeating: a straight comparison of GDP per person in dollars shows a difference between the two scenarios at the end of the century of $76,000. But climate change will cause net problems in both scenarios that reduce our prosperity or welfare. In the "fossil fuel" scenario, global warming will cause problems worth $11,000 per person each year, whereas in the cooler "green" scenario the costs will be much smaller at $3,000. So the difference in welfare, after we have deducted the climate damages, is now a slightly smaller $69,000 (which is the difference between $103,000 per person in the "green" pathway and the $172,000 in the "fossil fuel" pathway). Even after taking into account the climate damage, the extra welfare benefit of the "fossil fuel" pathway is still remarkably large.

Look at it another way. If we follow the "fossil fuel" pathway, the average welfare in 2076, after we have deducted the climate damage, will be $79,800 for every person in the world. Under the "green" scenario we won't reach that welfare before 2100. Choosing the latter means we're literally holding the world back by a generation. By the end of the century, the difference for the entire world population is a staggering $509 trillion *per year*.

This is the choice we are faced with. Just like the hypothetical choice in 1940 between two pathways, we have to ask, why would we choose a poorer world with fewer opportunities for our children and grandchildren?

WHAT I'M ARGUING HERE, that we should pick the higher-growth pathway, runs counter to the arguments of many climate change campaigners, and is especially at odds with the views of a vocal minority who believe that growth needs to be stopped altogether for the sake of the planet.

In late 2018, nearly 240 academics signed a letter declaring that economic growth is bad for the planet. They reject "sustainable" and "inclusive" growth as insufficient concepts, and argue that instead indicators of resource use and inequality should be given a higher priority than GDP in political decision making, that income should be capped, and that working time should be reduced. They and global supporters of the "no-growth economy" have held conferences in Mexico City, Mexico; Malmö, Sweden; and Brussels, Belgium.[11]

This is a movement that says more about the self-indulgence of rich country academics than anything else. But it is worth emphasizing how dangerous the concept is.

Obviously, the supporters of no-growth would argue that they do not merely want to stop economic growth, but want to encourage more education, health, and technology, especially for the world's poorest. But in a realistic world, less or no growth means that distributional fights become more vicious, and it seems naive to expect that ever-increasing shares of a small cake would go to the least powerful. What is clear is that of the UN's different projected futures, the low-growth scenarios are utterly unattractive, most especially for the world's poor.

Heeding the call of no-growth would likely mean we would have much fewer resources for health care, education, and technology. This outcome would trap us in a world with far fewer resources to further human welfare. In this scenario, the anti-growth campaigners would have succeeded in ensuring the world is only a small amount richer by the end of the century. Our dramatic progress in eradicating poverty would be halted. The world would be a much bleaker place—less healthy, less educated, less technologically advanced.

PICKING A LOWER-GDP PATHWAY does not just mean accepting dramatically lower welfare. The harsh reality is that reducing income also literally kills people.

Being richer means living a safer life. When you have a higher income, it allows you to buy more risk reduction, from safer cars, to bicycle helmets, to better medical care and more nutritious food. Researchers have looked at the impact of higher incomes and found that they lead to lower death rates, both because you can afford more health care and because you can afford to give your children better opportunities. All of this means that kids are more likely to survive and to thrive in adulthood. Having $69,000 worth of higher welfare for every person for every year can make you a lot safer. Using the estimates in the literature, it is likely that by the end of the century the "green" pathway will see three million more deaths each year compared to the "fossil fuel" pathway.[12]

We stand at a forked road, with different paths ahead of us. Our fear of climate change is pushing us down the "sustainability" pathway. This sounds like things we need: more inclusive development that respects environmental boundaries, shifting our emphasis away from economic growth, and reducing our consumption. Yet, the result is that humanity will be $500 trillion worse off, with more poor people, more inequality, less opportunity, and millions of more premature deaths than if we decide on the "fossil fuel" pathway. We will be choosing to make each person on the planet $69,000 worse off per year, even after taking the costs of climate change into account.

Following the "fossil fuel" pathway doesn't mean that we ignore climate policy. In section 4 we will talk about the right climate policies we should enact to ensure that we *also* tackle climate change. But it is more important for the bigger picture that we get all of our other policies right.

Notice what the UN clearly lays out is important for avoiding the bad scenario outcomes. It has almost nothing to do with climate. It is all about investing massively in education, health, and technology, especially for the world's poorest. Our preoccupation with climate change makes us forget these simple truths.

Choosing the right pathway will result in huge benefits from free trade, harnessing innovation and ingenuity, strong developments in human and social capital, and technological improvements, as well as more access to energy. If we choose the right pathway, we will leave our grandchildren and great-grandchildren with far greater opportunities. The benefits will especially help the world's poorest, as we will see in the next chapter.

HOW CLIMATE POLICY
HURTS THE POOR

"THE RICH POLLUTE, the Poor Suffer" runs a headline from the *Economist.* It's true: one of the insidious aspects of climate change is that although it mostly wasn't caused by the world's poor, it will hurt poorer countries more than richer countries. This is partly because the economies of poorer countries are more reliant on industries like agriculture that are vulnerable to climactic changes; it is partly because they tend to be in already warmer climates; and it is mostly because being poor means having less capacity to adapt.[1]

Climate change will cost most developed countries a few percent of GDP, which is a problem. But for the poorest countries and particularly for Africa, this cost can easily end up being much more, maybe 10 percent of GDP. That is a big problem.

This unfairness has long been used to argue for stringent climate policy. "Cut Carbon by Up to Third to Save Poor," the British newspaper the *Guardian* was already exhorting a decade ago. But the reality is that today's climate policy does little to tackle the climate-related problems of the world's poor.[2]

And it does virtually nothing to help with the vast array of nonclimate issues that the world's poor struggle with every day. In fact, climate policies often make life *worse* specifically for the poor. This is true both in rich countries like the United States and the United Kingdom, and in the poorest countries, especially those in Africa and Asia.

Today's blinkered climate policies are in fact putting the world's worst-off countries on a slower path to progress and prosperity. What are we doing in the name of climate?

WHEN DO PEOPLE in rich countries talk about the problems in the developing world? When there is a natural disaster. Typically, whatever the catastrophe is, it is blamed on climate change, and commentators are quick to make the case to cut more carbon dioxide. Unfortunately, this is an absurdly ineffective policy.[3]

Let's look at the devastating Typhoon Haiyan that hit the Philippine city of Tacloban in 2013, killing 2,700 people in the city, or 1 in 80 of its 218,000 inhabitants. Built on flat land in an area of the coast notorious for concentrating and amplifying storm surges, Tacloban had seen many disasters. In 1912, a hurricane of similar strength following almost the same path destroyed Tacloban and probably killed 6,000 people. But back then, Tacloban was much, much smaller, and it was estimated that 1 in 2 inhabitants were killed.[4]

Haiyan got global attention because it hit while the world's climate negotiators were gathered in Poland for their annual UN conference. The Philippine envoy spoke movingly of his grief. According to the *Guardian*, this made him the face of the UN climate talks and a "climate justice star." The diplomat vowed to fast until the world agreed to a solution involving dramatic carbon dioxide cuts. But will cutting carbon dioxide really help Tacloban?[5]

We saw earlier that climate change will likely mean fewer but stronger hurricanes. Cutting carbon dioxide will delay these effects slightly, but not reverse them. Even under the most dramatic climate policies, temperatures will still rise, only slightly more slowly. Hurricanes will still get stronger, only slightly less soon. Because carbon cuts will only slowly make any impact on rising temperatures, such action would help not one single person today or in the next decades. Cutting carbon even drastically would help future citizens of Tacloban only marginally in the sense that it would push stronger hurricanes a bit further into the future.

What if, instead of cutting carbon, we helped Tacloban escape widespread poverty, and got more people out from under flimsy, corrugated roofs? That would transform the lives of today's most vulnerable, making them more resilient, and of course it would dramatically improve the prospects for future generations.

Richer people are, unsurprisingly, more able to invest in adaptation to protect lives and valuables. One 2016 study shows that when income doubles for the average person in a poor community, fatalities from natural disasters will be cut by *more than a quarter*. What this tells us is that increasing incomes builds resilience. This is why the much poorer Tacloban of 1912 saw a forty times greater hurricane death toll (in relation to the city's population) than in 2013.[6]

Since the first climate summit in Rio de Janeiro in 1992, the world has talked incessantly about cutting carbon emissions. Imagine if in 1992 we instead had focused on cutting Tacloban's poverty. From 1992 to 2013, general economic progress in the Philippines tripled GDP per person; if we had focused *more* on poverty reduction, we might reasonably have quadrupled GDP by the time Typhoon Haiyan hit. This would have saved more than three hundred people's lives. Climate policies saved zero.

Putting it bluntly, choosing climate policies over growth policies doesn't just do nothing. It means more people die avoidable deaths.

As Tacloban and other vulnerable parts of the world get richer, inhabitants will be able to build better houses, transport, and infrastructure. Of course, this will raise the costs of property damage when hurricanes hit. Nonetheless, investment in these adaptation measures will mean that a doubling of average income will cause damage as a percentage of GDP to *decline* by 30 percent. Moreover, rising prosperity doesn't just mean more money for adaptation. It means more money for education, birth control, and investment. It means less malnutrition and lower infant mortality. Increasing prosperity makes life better for people in all sorts of ways.[7]

In contrast, cutting carbon emissions is generally an incredibly ineffective way to improve the lives of the world's poor. Helping them through carbon dioxide cuts, even very dramatic cuts, means the world will still see a significant temperature rise, only a slightly slower one. This means that whatever climate-related worry you try to address will still get worse; it's just that it will be slightly less bad than it could have been.

Worried about drought? Cutting carbon dioxide will not reduce drought risks—at best it will make drought increase slightly less quickly.

Worried about floods, heat waves, or the multitude of other possible climate impacts? The same thing happens. At best, you're slowing the problems slightly, at a great cost.

We need to open our eyes to a broader range of solutions, because we know that lifting people out of poverty has amazing impacts. Making people better off over a couple of decades means that any size storm they encounter will cause fewer deaths and less damage.

Lifting incomes significantly reduces the damage from any potential climate-change-caused increase in hurricanes, droughts, and floods. Moreover, there have always been natural disasters and there always will be, even if we stopped climate change altogether. Lifting incomes doesn't just help with the increase in natural disasters caused by global warming. It also helps reduce all the damage that would have happened anyway.[8]

FOR THE WORLD'S POOR, climate policies are often worse than ineffective. They are destructive.

Take nutrition. Climate activists often point out that higher temperatures will make more people hungry, so drastic carbon cuts are needed. But a comprehensive study published in 2018 in *Nature Climate Change* shows that strong global action to reduce climate change would cause far more hunger and food insecurity than climate change itself.[9]

The scientists used eight agricultural models covering the entire planet to analyze various scenarios for food security between now and 2050. Similar to how climate change models are used, one economic model might reflect the overly optimistic or pessimistic assumptions of a particular set of researchers, so it's best to look at a broad range of them, and at the averages.

Overall, these eight models give us reason for cheer: on average, they suggest that when disregarding global warming, predicted economic growth between now and 2050 would dramatically reduce the number of people starving, from almost eight hundred million today to two hundred million. But the models also allow us to see the effect of climate change. They show that rising temperatures could mean both lower food production and higher food prices. By 2050, this would put about twenty-four million extra people at risk of hunger who otherwise wouldn't have been.

That sounds like a good reason to aggressively push for climate policies to cut emissions. Strong carbon-cutting policies can indeed potentially alleviate some of this added vulnerability to hunger. If these climate policies had no cost at all, they would reduce the increases in temperature and save about fifteen million out of the twenty-four million extra people who would be starving because of climate change.

Unfortunately, as we have seen, aggressive action to reduce climate change is not costless. One of the biggest impacts is on energy prices, which hurt the poor the most. Indeed, the climate policy costs hitting not just households but also agricultural producers, food manufacturers, and the transport sector mean that food prices are predicted to climb by 110 percent by 2050. And that would have the total effect of forcing seventy-eight million people into starvation.

Which is to say that by insisting on cutting carbon through climate policies that push up food prices, instead of taking a broader view of how to best help people and the planet, we will have consigned fifty-four million *more* people to starvation. This is unconscionable.*

THE PARIS AGREEMENT is expensive and largely ineffective. It is also going to mean more people left in poverty. A 2019 study found that the massive cost of reducing emissions under the Paris Agreement will lead to an *increase* in global poverty (compared to what would otherwise be expected) of around 4 percent. The authors issue a stark warning that strong climate change policies could slow efforts to reduce poverty in poor countries.[10]

Let's look at the world's poorest continent, Africa, which will be particularly badly hit by global warming. The green and fossil fuel pathways

---

*In fact, I just told you the story without carbon dioxide fertilization, because this story sounds more "normal." If we include carbon dioxide fertilization (as we should), strong climate policy becomes an even more risible solution to hunger. The net impact of climate change will actually be *beneficial*, because carbon dioxide fertilization drives down food prices more than temperature drives them up. In total, global warming will mean eleven million *fewer* starving in 2050. Trying to limit climate change by cutting carbon dioxide will no longer solve a problem, but actually prevent a benefit. Of course, aggressive climate action will still increase prices and drive more people into hunger. Under these more realistic assumptions, climate policy is no longer a solution and will in total cause seventy-seven million *more* people to starve (Hasegawa et al. 2018, fig. S6).

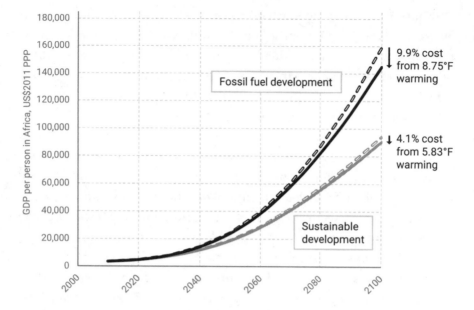

FIGURE 10.1 African GDP per person for the UN's sustainable and fossil-fuel-driven scenarios. The dashed lines show the GDP per person in dollars. Because higher temperatures also mean more climate damages, the solid lines show the welfare left after deducting the costs of climate damages.[11]

we saw in chapter 9 once again show two divergent futures (see figure 10.1). For Africa, instead of a twenty-fold increase in incomes from 2020 to 2100 in the green scenario, the fossil fuel pathway would deliver a thirty-three-fold increase. Of course, climate campaigners would immediately counter: "But climate change will devastate Africa."

Using a regionally defined climate cost model by William Nordhaus, we can roughly identify the total, negative impact from climate change on Africa. It is clear that Africa will be hurt the most of any region with global warming, and it will lose even more when temperatures go higher under the fossil fuel scenario. In 2100, Africa will lose 4.1 percent of its GDP to global warming if we follow the green pathway, and it will lose much more in the hotter world of the fossil fuel pathway, with a GDP reduction of 9.9 percent in 2100.[12]

But something else is also clear. While Africans will lose much less prosperity to climate change if we follow the "sustainability" ("green") pathway, they will be much better off *overall* if we follow the "fossil fuel" pathway: they will be more than thirty times more prosperous in 2100 than in 2020, even after accounting for climate damages, compared to "just" nineteen times better off in the "sustainability" pathway. By choosing the latter, we would be leaving each African $55,000 worse off every year.

The truth is that climate change plays a relatively small role in determining future well-being. It is clear that if we're motivated only by trying to reduce the impact of rising temperatures, we're literally ignoring the most important factors—such as education, health, technology, and access to plentiful energy—in Africa's future well-being.

WELL-MEANING CLIMATE POLICIES also have indirect impacts. Campaigners used to claim that rising temperatures would lead to more cases of malaria because there would be more warm places where mosquitoes would thrive. We don't often hear this concern anymore, thanks to tremendous success tackling malaria directly with insecticide-treated bed nets, indoor spraying, and treatment, which has halved the number of deaths over the last fifteen years. Imagine if we hadn't focused on these simple approaches, but instead put all our efforts into cutting carbon dioxide. We would still have 840,000 annual deaths, 400,000 more than today, while we waited for the miniscule effects of moderate carbon cuts to have any impact.[13]

Instead, lifting societies out of poverty is a much more important pathway to eradicating malaria. A study shows that when the average GDP per person reaches $3,100 per year, individuals have enough resources that most can afford to buy drugs to cure malaria, essentially ridding society of the disease altogether.[14]

Nonetheless, campaigners for carbon cuts continue to claim that the Paris Agreement is an important way to help tackle health issues like malaria. The World Meteorological Organization devoted a Paris Agreement–themed conference to health issues and proclaimed that implementation of the pact was key to improving health, including tackling malaria. Even the illustrious medical journal *Lancet* published a commentary by two doctors opposing the Trump administration's backpedaling on the Paris Agreement, citing climate change's effect on malaria as

one of the health impacts. Yet, the Paris Agreement will have no percep-
tible impact on malaria because it will lead to such small temperature
changes; in fact, it is very likely that its total impact will actually lead to
more malaria deaths.

Certainly, we know this to be the case for the world's previous climate
agreement, the Kyoto Protocol. It was found that had we implemented
that protocol for the rest of this century, it would through lowered temperatures have reduced the global number of malaria deaths by about
four hundred thousand over the century. Yet higher costs in the rich
countries implementing Kyoto would also have meant slightly slower
growth for trading partners like Africa, forcing them to linger in poverty
longer.[15]

The negative impact from lower growth would cause an extra six hundred thousand malaria deaths. This perversely leads to climate policies
making *more* poor people die from malaria because they delay the time
when nations get rich enough to see the final eradication of malaria.

We have been here before, implementing policies that do a huge
amount of harm in the name of climate change. Twenty years ago, a
craze for biofuels swept rich countries, with global production quintu-
pling in the first decade of this century. Biofuels are created with crops
rather than fossil fuels, so in principle they do not increase carbon dioxide emissions. Rich countries hurried to institute targets to promote
more biofuel use to help cut carbon emissions. The European Union led
the way, stipulating member states in 2003 pass legislation aimed at replacing 5.75 percent of all transport fossil fuels with biofuels by 2010. Developing nations, even those in the grip of famine, were pushed to grow
crops for ethanol instead of for food.[16]

This movement originally had the full-throated support of many
green groups that hailed the shift away from fossil fuels. Yet, the negative
impacts were much greater than most had anticipated. The charity ActionAid calculated that the amount of crops needed to fill an SUV's fuel
tank with biofuel would feed a child for an entire year, and every gallon
of biofuel wiped out forty meals.[17]

The huge growth in biofuel inevitably contributed to a reduction in
food and an increase in food prices: a confidential World Bank report

obtained by the *Guardian* found that biofuels had forced global food prices up by 75 percent. The results of the price hike were devastating. After food prices first spiked in 2008, the UN special envoy for the right to food, Olivier De Schutter, declared that a "silent tsunami" had pushed a hundred million people into poverty and thirty million into hunger. The World Bank subsequently estimated that between June and December 2010 an additional forty-four million people fell below the extreme poverty line because of food price hikes.[18]

Many environmental groups began to soften their support for biofuels or backtrack. Campaigners for stringent climate action were increasingly appalled. *Guardian* columnist and strong climate campaigner George Monbiot called the subsidies driving the biofuel industry's growth "a crime against humanity." Yet, by the point of the backtrack, vested agricultural interests had made the bad policies almost impossible to overturn.[19]

It seems that we have learned little from recent history, as we plow headlong into new policies that will similarly hurt the world's poor.

THE TRUTH IS that climate policies hurt the poor everywhere, even in countries like the United States, because higher energy prices have a disproportionate negative impact on the poor. Universally, poor people in well-off countries use much more of their limited resources paying for electricity and heating. One 2019 study showed that US low-income consumers spend 85 percent more on electricity as a percentage of total expenditure than high-income consumers. When climate policies lead to higher electricity prices, this harms the poor much more than the rich.[20]

This is one of the reasons why rich elites have no problem saying we should increase gas prices to $20 a gallon—they can easily afford it. Wealth also tends to be clustered in cities, where people drive much less. Of course, the struggling single mom in Huntington, West Virginia, has a very different experience.

Across the United States, many people are still struggling to pay their energy bills. The International Energy Agency estimates that 9 percent of Americans, or twenty-nine million, are "energy poor," spending more than 10 percent of their income on energy. This means poor Americans often literally have to forgo other basics in order to heat (or cool) their homes sufficiently.[21]

High energy prices literally kill people, as one 2019 study shows. Researchers looked at the natural experiment that happened around 2010, when fracking delivered a dramatic reduction in costs of natural gas. The massive increase in availability of natural gas drove down the price to heat homes. Cold homes are one of the leading causes of deaths in winter: there is a strong connection between low indoor temperatures and increased risk of strokes, heart attacks, and respiratory diseases. So reducing the price of heat saves lives. The study estimates that these new lower energy prices save about eleven thousand Americans from dying in winter *every year*.[22]

If climate policy is to work, it will drive up prices of energy to reduce consumption. So a climate policy reversing the price reduction due to fracking will drive energy prices back up. People will be less able to heat their homes, and the consequent death rate will go back up.[23]

Higher energy prices also hit the poor the hardest in other rich countries. In the United Kingdom, rich households spend 3 percent of their income on energy, whereas poor households spend 10 percent. Campaigners celebrate that the UK has seen a reduction in electricity consumption, but this is mostly because costs have soared. Ever more stringent climate legislation made the cost of UK electricity increase by 62 percent in real terms over a decade, during a period in which average incomes increased just 4 percent. Not surprisingly, it is the poor who have done almost all of the cutting back, reducing their consumption much more than average, while the richest cut almost none of their much higher electricity use.[24]

Higher energy prices hit especially the elderly, who often have to survive on low incomes. One poll in the UK showed that a third of elderly people leave at least part of their homes cold, and two-thirds say they wear extra layers of clothing as a result.[25]

IN THE WORLD'S poorest countries, the insistence that we should limit future access to fossil fuels has an even more dramatic cost, because

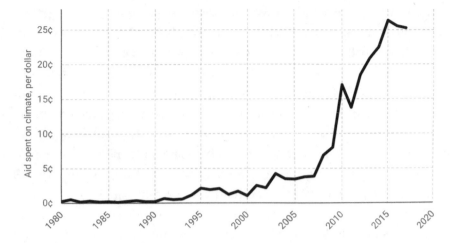

FIGURE 10.2 For every dollar spent on aid to the developing world, the number of cents that go to climate purposes.[26]

cheap and accessible power is life saving, and one of the best ways out of poverty.

Yet, many development organizations with a primary goal of helping the poor, including most notably the World Bank, direct more and more of their help to what is known as "climate-related development finance" or "climate aid." These are nebulous terms, but often what they mean is more money for projects that build solar-energy microgrids and an explicit refusal to fund any fossil-fuel-related projects like new coal power plants.

Efforts to calculate just how much of global development aid goes toward climate-related projects (see figure 10.2) make it very clear that the percentage increased drastically from about 2005.

Climate aid does only a little good. Remember the Indian village of Dharnai (chapter 7), where the villagers wanted "real" electricity that could power their refrigerators, stoves, and also the lights in the morning? In general, studies show that aid spent on solar panels and other climate initiatives generates some of the smallest benefits compared to other aid investments in nutrition, health, gender, education, and infrastructure.[27]

The International Energy Agency expects that 195 million people will have access to energy for the first time in the next decade. That is great.

But they will get very little power, mostly because the focus is on getting them off-grid solar rather than on-grid, mostly fossil-fuel-based energy. Indeed, they will on average get just 170 kilowatt-hours per year. That is half of what one US flat-screen TV uses in a year. Consider what your life would be like with access to nothing but 170 kilowatt-hours. It is not enough to power a factory or a farm, so it cannot reduce poverty or create jobs. It won't even be enough to hook up a stove and heater to combat the fire-created smoke and pollution that claims 1.6 million lives annually. It's not even enough to keep a refrigerator running. For the world's poor, this level of energy access is hardly transformational.[28]

In contrast, a study in Bangladesh has shown that grid electrification, which in Bangladesh and much of the world mostly means using fossil fuels, has significant positive impacts on household income, expenditure, and education. It means households newly introduced to grid electrification experience on average a 21 percent jump in income, and over a period of nine years a 13 percent reduction in poverty.[29]

Unfortunately, many of our political leaders have decided that when spending on aid for the world's poorest, climate change concerns trump poverty. That calculation is wrong, and is undermining the claim to helping the world's poor.

In one 2016 study looking at energy options for Bangladesh, researchers found that building more coal-fired power plants would generate global climate damage costing around $592 million over the next fifteen years, but the benefits from electrification would be almost five hundred times greater at $258 billion, equivalent to more than an entire year of the nation's GDP. By 2030, the average Bangladeshi would be 16 percent better off.[30]

Denying Bangladesh this benefit in the name of reducing the impact of climate change is fantastically arrogant: for every 23¢ of global climate damage that we could avoid, we are asking Bangladesh to forego $100 of prosperity and development. And Bangladesh is a nation where energy shortages cost an estimated 0.5 percent of GDP, and around twenty-one million people live in extreme poverty.[31]

From the comfort of the World Economic Forum's 2017 annual meeting in Davos, Switzerland, former US vice president Al Gore tut-tutted about plans to build coal-fired power plants in Bangladesh. But Bangladeshi prime minister Sheikh Hasina slapped that down, pointing out: "If you cannot develop the economic conditions of your people, then how will you save our people? We have to ensure the food security; we have to give them job opportunity."[32]

Sheikh Hasina is right. Choosing expensive carbon-cutting policies or insisting on green development approaches might seem like an easy choice for the world's elite in Washington, DC, or Paris, France, but the burden of these choices falls unfairly on the world's poor, and especially on those living in abject poverty. They need more energy, not moralizing from the West.

It is perverse to hear rich people piously claim that we should help the world's poor by cutting carbon dioxide to make their future slightly less worse, when we have huge opportunities to make their lives much better, much more quickly, and much more effectively.

IF YOU ARE POOR, you will be less able to build a strong house to withstand hurricane-level winds or flooding. You will be less able to buy an air conditioner to adapt to rising temperatures. But if you are poor, you will also lack resources to save your kids from easily curable diseases, to keep them well nourished, to educate them, to house them in a safe neighborhood, to ensure they are not exposed to air pollution—the list goes on.

The truth is that poverty sucks. Exposure to natural disasters caused by climate change is just one of many ways that it sucks, and it happens to be one of the ways where we can help the least even when spending trillions. And when we spend lots of resources on climate, there is less money left to tackle all the other problems facing the poor, problems where we can help much more.

Today there are about 650 million extremely poor people in the world. As a thought experiment, let's consider what it would cost to lift all these people out of poverty. (As a caveat, let's just remember that lifting people from, say, 90¢ a day to just above $1.90 is great, but they are still pretty darn poor—and actually distributing the money would be fantastically expensive.) It turns out that the theoretical cost to lift *everyone* on the planet out of extreme poverty would be less than $100 billion per year.[33]

Compare this to our current trajectory: we've committed to spending $1 *trillion* to $2 *trillion* a year just on the almost entirely ineffective Paris Agreement. Every *month* the cost will be the same as the amount that could lift *everyone* from extreme poverty. This strikes me as obscene. As rich countries commit to going carbon neutral, the costs will escalate to tens of trillions of dollars per year, to make a small temperature change in a century's time. Just a couple of days of these new, higher costs could transform the world by ending extreme poverty entirely.

CLIMATE ALARMISM TOO often leads us to policies that while well intentioned, crowd out much more effective ways of helping people. It comes down to this: when we see a malnourished child or a town hit by a hurricane and seriously suggest that we should make lives better by cutting a ton of carbon dioxide, we are not actually trying to do good, but rather imposing our own priorities on people who have little power to assert their own. It has become too easy to believe that policies aimed at cutting carbon are the answer to everything. They're not, and we need to stop campaigning for and enacting policies that will have the world's poorest paying for our mistakes.

Thirty years of climate policy have failed to rein in temperature rises or reduce carbon intensity—the amount of carbon dioxide we emit for every unit of energy produced. All the well-meaning personal actions undertaken in the rich world, like buying electric cars or becoming vegetarian, amount to little more than gestures. Far more importantly, the Paris Agreement has put us collectively on a pathway toward incurring gigantic costs, especially for the poorest, with next to no climate benefit.

Thus far, humanity has excelled at showing how *not* to fix the climate. We have spent three decades trying the same, deeply broken approach, over and over. Politicians lurch from one climate summit to the next, with climate campaigners urging them on to make even more ludicrous promises. Enough is enough.

There are smart policies that can tackle climate change and help make the world better. These are what we'll turn to in the next section.

HOW TO FIX CLIMATE CHANGE

CARBON TAX:
THE MARKET-BASED SOLUTION

IT IS TIME to admit that the current approach to fixing climate change is not working. More protests and promises, more solar panels and wind turbines, more subsidies—all of these just tinker at the edges. We need to resist the shortsighted call from campaigners, green lobbyists, and populist politicians to double down and make even bigger carbon-cutting promises. After thirty years of failure, we need to say enough is enough.

The good news is that there are smarter, more effective policy options. We should find and implement ways to cut emissions effectively, both so the cuts cost as little as possible and so they do more good than harm. But we also need to harness the best of humanity—the innovative spirit and creativity that can generate new solutions.

In this section, I will outline five key policies that most people can agree on, because they are a smart and effective way to tackle climate change.

THE FIRST WAY to tackle climate change is to effectively implement a tax on carbon dioxide emissions, usually just called a "carbon tax." I want to first focus on why we should have a carbon tax, and then at what level it should be set.

Carbon taxes can aggressively reduce emissions and thereby limit global warming's most damaging effects at fairly low cost. This is not a controversial idea. Most economists agree that the most effective way to reduce the worst damage of climate change is to levy a tax on carbon dioxide emissions. These emissions are simply a by-product of burning fossil fuels to achieve productive things like harvesting crops, pumping water, and keeping medicine cool. The benefits accrue to the person or company making the emissions, and to the consumers of the products they make, but the negative effects (carbon dioxide emissions and their impact on temperatures) are spread over a large population.

This is a classic example of a "market failure": one person gets the benefits, someone else gets most of the "disbenefits." On a much smaller scale, imagine that you burn a fire in your home's fireplace; it creates a lot of annoying, dirty smoke that makes you cough, and soot that covers everything. If you build a tall smokestack, you've solved the problem, for you, and shared it with a much larger population. The problem itself, though, hasn't been solved.

There are various ways to tackle the market failure at the heart of climate change. The reason a market works in the first place is that prices are supposed to deliver all the information we need. When you sit down in a restaurant to order wild-caught salmon, you don't need to know whether this is the right season for salmon fishing in Alaska, whether the seas have been rough or the trip to the harbor was particularly difficult, or whether the chef is having financial problems. You simply look at the price and make your choice based on that, which should already incorporate all these factors into one number. The market failure comes when some costs do not show up in the final price; for instance, the carbon dioxide emitted by the fishing boat and the planes and trucks that brought the fish to market, all of which will cause some climate problems.

One solution to the market failure could be to ban the use of fossil fuels. That would be terribly inefficient, since it means that humanity would forgo any benefit, no matter how large, from fossil fuels in order to avoid the negative impact from climate change.

Another solution would be a set of regulations: the government could decide how much carbon dioxide the boat is allowed to emit, restrict flights and what cargo can be flown where, and control how far trucks are allowed to drive to deliver the salmon. But to control all these emissions of carbon dioxide, governments would essentially need to regulate every minute part of the economy. Every regulation bears a cost: it ties up businesses and citizens in a little more red tape. At the required scale, the price tag would be huge.

Instead, economists point out that we can correct the market failure with relative ease by putting a cost on the carbon dioxide that is emitted to make sure the price of any product or service, including the fish we are ordering at a restaurant, incorporates the climate damage it will cause. This is essentially what a carbon tax does. It forces you to take into account the climate disbenefits that your purchase is responsible for, so you can weigh these against the benefit.

Let's consider what happens to our salmon. Carbon emissions occur at each stage of the process of getting the fish from the ocean to the table. With a carbon tax, the fishing company would pay a little extra for fuel. The transport company flying the salmon from Alaska might have to pay quite a bit extra. Even sautéing the salmon in the kitchen will add a tiny extra cost because the restaurant pays slightly higher gas prices. The carbon tax not only encourages each participant in the market to be more efficient, but the costs are passed on down the line. Eventually, the extra price that you the consumer pay in the restaurant reflects all these additional carbon damages.

Now, if you really like fresh Alaskan salmon, you will just pay the additional cost. But salmon frozen and transported by ship instead of by plane will be responsible for much less carbon dioxide emissions and thus incur a much lower additional tax, so perhaps to save money you will buy frozen fillets from the supermarket and cook them yourself at home. In essence, the carbon dioxide tax has helped you discover that your enjoyment of fresh salmon does not outweigh the negative effect of the extra carbon dioxide emissions, and it has helped you change your choice.

If the carbon tax is set at the right level, it corrects the market failure so that the prices now reflect not just how difficult it was to catch and transport your salmon, but also its emissions along the way. Indeed, it doesn't just show consumers which products are carbon intensive and should be used more sparingly, but it helps energy producers move toward lower carbon dioxide emissions (perhaps through more reliance on solar and wind energy), and it encourages innovators to come up with new, lower-carbon processes and products.[1]

Unfortunately, another consequence of this tax is that it makes everyone slightly poorer than they would have been. It forces people and companies to use more expensive technologies and fuels than they otherwise would have done. This is crucial to our discussion of how to set a carbon tax. Emissions damage the climate, creating costs over the long term. Climate policy, through a carbon tax, reduces those costs but also causes its own damage to the economy. Having a climate policy means that we have to pay both costs, so we have to work out at what level a carbon tax should be set in order to minimize the impact of both.[2]

LIKE ANY OTHER TAX, a carbon tax can be set at any level that a government chooses. How does a government decide on the right carbon tax? There are costs to ignoring climate change and having no carbon tax, but there are also costs from exaggerating the impact of climate change and having a massive tax.

If we don't implement a carbon tax, we will bear the full brunt of climate change over the coming centuries. In chapter 7, we saw that the cost in the year 2100 will be about 3.6 percent of GDP, and there will be costs in the eighty years leading up to 2100 and far beyond into the future. Using Nordhaus's model, adding up all the costs throughout the next five centuries as if we had to pay them today, the bill would be $140 trillion. That is the top bar in figure 11.1, which shows the total global cost from climate change damages we would see without any climate policy.[3]

We can most effectively reduce the damages from climate change by having one carbon tax across all countries and goods. That way, producers can't just shift their emissions to a factory in another country, and everyone faces the same costs. Achieving a uniform global standard means all countries need to set the same carbon tax for all their emissions. Of course, this will be challenging to achieve (and we will discuss what happens if we fail, below), but one global carbon tax is by far the best approach. It will be much more efficient than if countries all go it alone and introduce varying levels of taxes. Moreover, using a single, global carbon tax is how almost all climate economic models look at the world.

It's important to note that this isn't a tax we should set at one level forever. A carbon tax that starts out low and increases over time will be

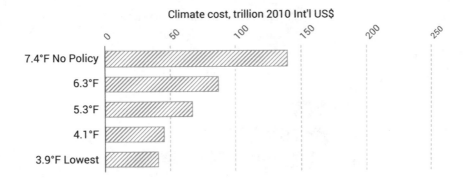

FIGURE 11.1 Total global climate change damage costs over next five hundred years for different temperature rises by 2100, using Nordhaus's Dynamic Integrated Climate-Economy model (DICE).[4]

able to cut more emissions at lower cost. That is both because technological developments will make future carbon cuts cheaper (this innovation is included in Nordhaus's model) and because stronger carbon cuts are more valuable in the future when temperatures and damages are higher.[5]

If we aim to reduce the temperature rise to 6.3°F in 2100, we can use Nordhaus's climate-economic model to find the most effective, increasing global carbon tax to achieve this goal. As shown in figure 11.1, successfully implemented, this tax would cut the total climate change damage costs from $140 trillion to $87 trillion.

But let's say that we instead aim for a 5.3°F temperature rise in 2100. Then we will need a higher, increasing carbon tax to achieve this lower temperature rise. If the tax is successfully implemented, we would cut the cost of climate change damage to $67 trillion.

As we cut the temperature more and more with ever higher, increasing carbon taxes, the climate change damage cost declines more and more. If we reduce the temperature rise to a very low 3.9°F, then damage cost would be cut by more than two-thirds, to $40 trillion. This obviously seems to suggest that we should simply cut temperature as much as possible, because then the climate change damage costs are the lowest. In fact, this is where most climate policy debate begins and ends: temperature rises are bad, therefore we must cut them as much as possible. But, remember that there is a second cost: cutting temperatures through a carbon tax reduces economic output. Let's call this the climate change policy cost.

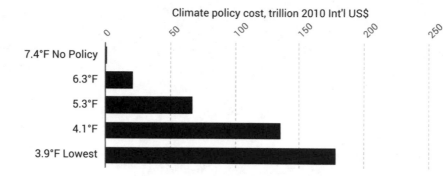

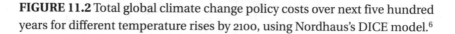

FIGURE 11.2 Total global climate change policy costs over next five hundred years for different temperature rises by 2100, using Nordhaus's DICE model.[6]

For the climate change debate to become more rational and pragmatic, we need to realize that we have to pay *both* costs: the climate change impact cost *and* the climate change policy cost. Making the temperature rise smaller means making the climate change policy cost bigger.

In figure 11.2, we see what happens to the climate change policy cost under different scenarios. Let's say that from 2010 we enact zero new climate change policies. This means that we allow for the small number of climate policies enacted around the world before 2010, but we imagine that no new policies will be introduced by any government to achieve any of the many climate promises made in Paris and elsewhere. The climate-economic model of William Nordhaus shows that doing so implies a temperature rise of 7.4°F by 2100, with $0 in new climate change policy cost, as seen in the top bar of the graph.[7]

However, if we want to limit global temperature rises to 6.3°F by 2100, the model shows that we can most effectively achieve that with a global carbon tax in 2020 of $36 per ton of carbon dioxide, increasing to $270 per ton by the end of the century. This will mean more expensive energy, which in total will result in costs of $21 trillion. This is the climate change policy cost shown by the second bar of figure 11.2.[8]

As we make climate change policy more ambitious, these climate change policy costs increase. Keeping temperature rises to 3.9°F by 2100 requires a global carbon tax that almost immediately rises to a high $500

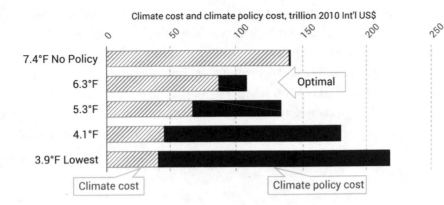

FIGURE 11.3 Total global climate change impact cost and climate change policy costs over the next five hundred years for different temperature rises by 2100, using Nordhaus's DICE model. Climate policy is, implausibly, expected to be efficiently implemented across all countries and centuries. Optimal policy (with the lowest total cost) indicated.[9]

per ton of carbon dioxide. How would that tax level affect your life? Well, you could expect to pay another $4.50 for every gallon of gasoline, and there would be similar impacts on all other items containing energy. And given that it is a global tax, everyone in the world, even in poor countries, would have to pay the same extra costs. At those kinds of prices, carbon dioxide emissions would essentially be eradicated within a decade. Not surprisingly, this policy will be very expensive at $177 trillion. (Notice we don't include the cost of the 3.6°F (2°C) target here, because it is likely impossible to reach that target with realistic technologies.)[10]

But because in setting climate change policy we need to take into account *both* climate costs and climate policy costs, let's now put both together, in figure 11.3. The hatched bars on the left show the climate cost, which is declining with more stringent climate policy. The black bars on the right show the cost of the climate policy, which increases as our policies become more stringent.

By adding the costs together at various levels, we can see the *optimal* climate change policy: the point where the *combined* cost of climate change damage and climate change policy impact is lowest. This is where the overall benefits to society are the greatest. If we cut temperatures beyond this point, the cost of climate change policy rises faster than the avoided cost of climate damages. If we cut carbon emissions below this point, climate change damage costs rise faster than the avoided climate change policy costs.

In Nobel laureate William Nordhaus's economic model, this optimal point is to be found if we keep temperature rises to 6.3°F in 2100. At 6.3°F, the climate change damage costs would be $87 trillion, and the climate change policy costs would be $21 trillion, leading to an overall cost of $108 trillion: this is the smallest possible combined price tag the world can achieve.[11]

MANY CAMPAIGNERS BELIEVE that allowing temperature rises of 6.3°F is not nearly ambitious enough. But that is only because it has become commonplace to talk about incredibly costly or even impossible temperature cap targets like 3.6°F (2°C) or 2.7°F (1.5°C) without any acknowledgment of their costs or plausibility.

Look at the emission reductions implied by the 6.3°F and the 3.9°F policy targets in figure 11.4. On the left-hand side, we see emissions over the century. The black line shows how emissions keep increasing if we do nothing. The gray line shows what would happen to emissions with the optimal carbon tax identified above. The dashed line shows emissions under climate campaigners' favored scenario.

Notice how the optimal tax immediately reduces emissions as the entire world adopts a common carbon tax in 2020. This would already constitute an amazing achievement. Then emissions increase somewhat as poor countries get richer and emit more, but just a couple of decades later emissions begin to decline, ending with less than one-fifth of the level expected by century's end.

If the world were to successfully adopt this optimal carbon tax, it would have a transformative impact: in 2100, the annual global emissions would be cut by more than 80 percent compared to what they would otherwise have been. Even though the cuts in the early decades are smaller, the total reduction over the entire century will be half of what we otherwise expected.[12]

Nonetheless, the impact on temperature of this intense emissions reduction resulting from the optimal carbon tax, as seen on the right-hand

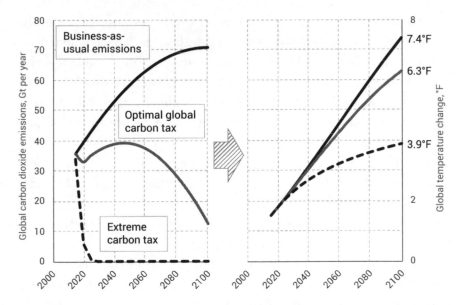

FIGURE 11.4 On the left, three possible emissions scenarios for the rest of the century. The black line is if we do nothing. The gray line shows the optimal global carbon tax, discussed above. The dashed line shows the favorite scenario of climate campaigners, essentially eradicating carbon dioxide emissions by 2030. On the right are the temperature increases for these three emission scenarios.[13]

side, is rather small. It reduces the temperature in 2100 from 7.4°F to 6.3°F. This is because the planet's climate has a huge amount of inertia built in, which translates even fairly dramatic reductions in emissions into rather small temperature changes. Even as we reduce our annual emissions, the total amount of carbon dioxide increases, just more slowly. Remember the metaphor of the bathtub? Even if we pour in less water, the water level still increases.

Moreover, achieving this temperature reduction requires the entire world to implement a rather steep carbon tax of $270 per ton by the end of the century. That equates to your average American motorist paying an extra $2.40 for every gallon of gasoline, and similarly high costs on all other uses of energy from fossil fuels. And even more importantly, it means every Chinese, Venezuelan, and Nigerian will also have to pay $2.40 more per gallon of gasoline and will face similar cost hikes on all other fossil fuels. This will be a tough sell.

Yet, this is nothing compared to what would be required under the most ambitious 3.9°F climate policy: essentially outlawing global fossil fuel emissions in a decade. First, it is practically inconceivable that this policy could be actually implemented throughout the world. Imagine telling Americans that they can have no gasoline for their cars by 2030, no trucks or trains running on fossil fuels, no power from coal or gas, no plastics, cement, fertilizer, and so on. Now try to imagine telling the Chinese to shut down 80 percent of their power stations. Or telling India and Africa they will have to forego having plentiful, cheap, and reliable power in the first place.

Second, even if we somehow managed this feat, the costs would be spectacular. The temperature will keep going up, though not quite as much. And the total costs to society, from climate change damage and climate change policy, would double (from the optimal cost of $108 trillion to $218 trillion, as seen in figure 11.3).

When politicians and campaigners talk about extremely drastic climate policies, they don't acknowledge, and perhaps don't even realize, that those policies have a cost to society vastly greater than the costs of the damage they are trying to avoid. We want to avoid climate change damage. But implementing a climate change policy also has costs, and the more ambitious the policy, the higher the costs. The policy that will leave the world best off is the one where we avoid the worst climate change damages *and* avoid the worst climate policy costs.

UNFORTUNATELY, THE OPTIMAL climate policy requires a globally coordinated carbon tax, and that is possible only in a fairy-tale world. It doesn't ever happen in real life. It would require politicians to set rational, ever more costly carbon taxes covering every person on the planet, through the next eighty years. Just in the United States, such consistency would be impossible. Over the next eighty years, the US will have forty different congresses and could have twenty different presidents.

Taxes are not set, nor adjusted, with global coordination and precision. Instead, they are set and unset nationally according to political opportunism, groundswells of opinion, and political setbacks. Look at France, for example, where President Emmanuel Macron backed down in

2018 on a 13¢ per gallon "green" fuel tax in the face of Yellow Vest protests. Look at President Trump in the United States, President Jair Bolsonaro in Brazil, and the many other politicians backtracking on climate policies.[14]

It beggars belief to think that climate change policy will follow an optimal path. The chances are basically zero that there will be one uniform, steadily increasing global carbon tax carefully coordinated among China, India, the United States, Iran, Saudi Arabia, the European Union, and every other nation over the next eighty years and beyond. The real world will see much muddling through, with some countries setting carbon taxes too high, others setting them too low, and all adjusting them up and down according to local events and political pressures. This will mean that any actual carbon tax will become more expensive than the theoretically pure model with a single, global carbon tax.

We have a good sense of how much more expensive climate policies could get. First, we know that the cheapest carbon cuts can be made in the developing world. Unfortunately, most actual carbon cuts are made in the rich world. It turns out that this alone makes the cost double, because we insist on cutting where the cost is highest.[15]

Second, politicians often pick ineffective climate policies. As we saw in chapter 8, the European Union has seen its climate change policy cost *double* from the most effective level it could have been, because politicians didn't make the most efficient choices and couldn't help but dish out favors to some industries over others. Similarly, much climate policy in many US states, including New York, California, and Hawaii, is based on the renewable portfolio standard, a regulation that according to economists makes it at least *twice* as expensive as the most efficient policy would have been.[16]

Thus, it is reasonable to expect that the real cost of climate change policy will be *at least* twice as expensive as the model we used in figure 11.2 and figure 11.3 suggested. (If anything, this is far too low because the evidence implies a doubling of costs *both* for coordination between the rich and poor world and for bad implementation in places like the EU and California, thus suggesting that the real amount could be at least four times the most efficient cost.)

To adjust for the real world, we can run the numbers through William Nordhaus's model again, but this time allow for policy costing twice as

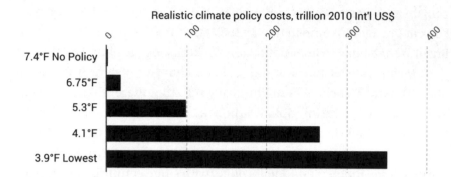

Realistic climate policy costs, trillion 2010 Int'l US$

FIGURE 11.5 Realistic global climate change policy costs over the next five hundred years for different temperature rises by 2100, reflecting additional cost due to political inefficiency, using Nordhaus's DICE model.[17]

much. Just as before, the costs of climate change damage still decrease as we reach lower and lower temperatures by 2100. Figure 11.5 shows climate change *policy* costs when we account for inefficiency. If we do no new climate policy, the new climate policy cost is still zero. But the climate change policy costs increase much faster when we acknowledge that they are inefficiently implemented and poorly coordinated. Now, every other amount is about twice as big. The plausible cost for achieving the most ambitious target of keeping temperature rises to 3.9°F by 2100 is a whopping $350 trillion.

Now let's again bring this all together in figure 11.6, this time including both climate change damage costs *and* realistic climate change policy costs. The optimal temperature rise is now 6.75°F (versus 6.3°F in figure 11.3) because realistic climate change policies are more expensive. The far costlier climate change policies required to achieve lower tempera-tures would be incredibly onerous. They would force the world to pay close to $400 trillion to achieve a temperature rise of 3.9°F.

In a realistic world starting from the close to no climate policy that we have today, careful climate policies can save us $18 trillion (the difference between $140 trillion in no-policy costs and the $122 trillion from the op-timal policy). Reckless policies, by contrast, could cost us $250 trillion dollars more (the difference between the no-policy cost of $140 trillion and the $390 trillion cost for the 3.9°F policy). Despite everything we are

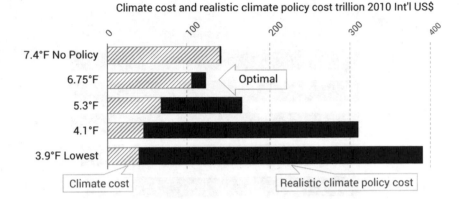

FIGURE 11.6 Total global climate damage costs and total realistic climate change policy costs over the next five hundred years for different temperature rises by 2100, using Nordhaus's DICE model. Optimal policy (with the lowest total cost) indicated.[18]

told, climate policies have a small upside (that we should exploit) and a potentially very large downside.

LET'S ASSUME THAT we all agree that our common goal is to create the best possible world for the generations that succeed us; that is, to create the maximum possible welfare for subsequent generations.

In this chapter so far, we have calculated the welfare loss from climate change damage, and the welfare loss from climate change policies, and we have looked for the sweet spot where we achieve the smallest welfare loss from both combined. But Nordhaus's model actually lets us take an even broader view and estimate the total amount of welfare available to society over the next five centuries under different climate policies. This is a much bigger picture, and puts the impacts of damages from climate change and damages from carbon taxes into perspective, because climate change is just one part of our future.

It may seem very abstract, but future welfare essentially represents what humanity gets to enjoy—this is all the "good" that we can make sure is available to future generations. We have a responsibility to maximize this good, to help future generations as much as possible.

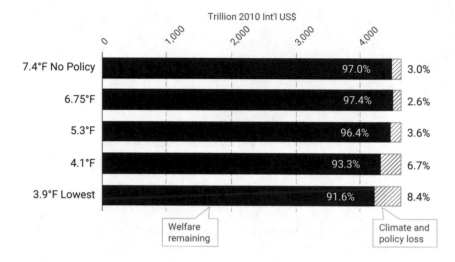

FIGURE 11.7 Value of total global GDP over the next five centuries. The hatched cost is the loss due to climate damage and climate policy; the rest in black represents remaining human welfare.[19]

In figure 11.7, the black bars denote the total welfare available to society over the next five centuries. It is the total GDP "left over" when we have both endured the cost of climate damages and paid the cost of climate change policies (the hatched area).

One important point to note is that humanity's welfare (the black) is large, no matter which climate policy we end up choosing. Even measured over the next five centuries, climate change damages and climate change policy costs will *not* do more than nibble away at the world's GDP. In other words, climate change and climate policy are responsible for only a smaller part of our future welfare.

It is likely that total GDP across the next five centuries will add up to $4,629 trillion—a huge figure. Damages from climate change and the costs of climate change policy will set us back. If we end up doing nothing to tackle climate change and reach a 7.4°F temperature rise by the end of the century, climate change damage will be significant at $140 trillion, or about 3 percent of total future GDP. Nonetheless, we will have 97 percent of all resources, or almost $4,500 trillion, left for health, education, food, and all the other opportunities that build humanity's welfare.

If we manage to implement a smart but realistic increasing carbon tax across the centuries, we can do better. With real-world assumptions, carbon taxes will be somewhat coordinated but not perfectly aligned, and they will rise across the century in all nations, somewhat haphazardly. The average global temperature rise will reach a slightly lower 6.75°F by the end of the century. The total costs of climate change policy and damage will be around $122 trillion, or 2.6 percent of the total GDP. This means that humanity will have 97.4 percent, or slightly more than $4,500 trillion, of the world's future GDP left for welfare.

Thus, the damage to GDP ends up 0.4 percent less than that in our scenario with no climate change policy. *A realistic, moderate and increasing carbon tax policy can definitely make a net benefit for society worth $18 trillion, or 0.4 percent of GDP.*

But what happens if we aim to enact much more ambitious climate targets, through much higher carbon taxes? The total costs increase dramatically. Trying to keep the temperature rise to 3.9°F in 2100 is still *less* ambitious than the popular idea enshrined in the Paris Agreement of keeping temperature rises under 3.6°F (2°C). Yet, the total loss to humanity from climate change and policy damage would reach a staggering $391 trillion, or 8.4 percent, of GDP.

Compared to the $140 trillion, or 3 percent, loss from doing nothing, doing too much is far worse. In an attempt to ameliorate climate change, we will end up avoiding more of the climate damage costs, but saddle the world with climate policies so expensive that the total costs almost triple. The cure is much worse than the disease.

CARBON TAXES ARE a good idea, and should be implemented as efficiently as possible. We need to be realistic, however, about what they can and cannot do. A modest, increasing carbon tax can be a very good idea if globally embraced, even if imperfectly delivered. It can ensure that we cut the most damaging of our emissions at low costs. Yet, the impact of optimal policy will be relatively modest, reducing temperature rises from 7.4°F to 6.75°F, or by about 0.65°F. The benefit will be worth around 0.4 percent of GDP.[20]

Realistically, a smart carbon tax will make an impact, but it can be only a small part of our solution to climate change, simply because going further means enduring more costs from taxes than the avoided costs of climate change.

We should implement a smart, global tax. But if we want to deal with more of the climate damages, a carbon tax is far from enough: we need to look at other approaches.

INNOVATION:
WHAT IS NEEDED MOST

THE BIGGEST PROBLEM we face when we confront global warming is that fossil fuels produce carbon dioxide. We need to reduce emissions to limit warming, but we have no easy and cheap alternatives to fossil fuels.

That's a big problem, but it is not humanity's first big problem. When we look back in time, we seldom fixed big problems by telling people to live with less of everything they wanted. And for a good reason: it is a hard sell. Instead, we typically managed to solve big problems through innovation.

We can draw lessons from the time we transitioned our global economy from one source of light to another. From the 1700s through the mid-1800s, whale oil provided light to the United States and much of the Western world. The US dominated the slaughter of whales, a barbaric practice by today's standards. At its peak, whaling employed seventy thousand people and was the fifth-largest industry in the United States. Producing millions of gallons of whale oil each year, the industry was widely seen as unassailable, offering a brighter and cleaner-burning option to cheaper, less safe alternatives such as lard oil and camphene. It was hard to imagine that people would ever agree to live without whale oil, because that would have meant going backward toward a sootier, dimmer past.[1]

In those days, of course, there was no environmental movement to speak of. But one wonders if the whalers, finding each year that they needed to go farther from Nantucket Island to kill whales, ever asked themselves: "What will happen when the whales run out?"

The Western world was reliant on slaughtering whales for quality lighting, yet we never did hunt whales to extinction. Why? We found alternative technologies. First, kerosene from petroleum replaced whale oil as a source of lighting. And we didn't run out of kerosene, either: electricity supplanted it because it was a superior way to light our planet.

Exhortations to stop using whale oil, to turn down the lights, or go back to old, heavily polluting practices were not what saved the whales. New technology did.

We have consistently underestimated our capacity for innovation. Before the turn of the last century, there was a fear that all of London would be covered with horse manure because of the ever-increasing use of horse-drawn carriages. Innovation gave us the car. Today, eight million people live in London, and there's no horse manure to be seen on its crowded thoroughfares.[2]

When automobiles proved to be massive polluters and Los Angeles was filling up with smog in the 1960s, one solution would have been to tell everyone to stop driving. Not only would it have been unfeasible to try to make people stay at home, but it would also have been phenomenally ineffective. Instead, the invention of the catalytic converter made it possible to have many more people drive many more cars and still have dramatically better air quality.

Indeed, many forecasted catastrophes throughout human history have been avoided because of innovation and technological development. Consider the challenge of hunger in the 1960s and 1970s. The biggest fear then was Asia's inability to feed itself, because it wasn't growing enough food for a fast-growing population. One of the best-selling books of the day, *The Population Bomb*, declared straight-out that "the battle to feed humanity is over. In the 1970s hundreds of millions of people will starve to death in spite of any [policies] embarked upon now."[3]

The belief was that India was doomed. Conventional wisdom was that the nation's 1967 grain output of barely a hundred million tons could not possibly increase. It was expected that two hundred million more children would be born by 1980, so disaster on a massive scale appeared unavoidable.[4]

Along came a determined scientist, Norman Borlaug. Instead of handwringing, he aggressively helped innovate new, better varieties of wheat, rice, and maize. These new dwarf plants differed from existing varieties because less energy was used in the stalk and more energy went into the grains themselves. As a result, grain yields increased spectacularly, food prices dropped, and Borlaug's "green revolution" probably saved a billion people from starvation.[5]

By 1980, India had seen a 47 percent increase in grain production, while its population increased 34 percent. Calories available began to increase, not decrease. Today, India produces 328 percent as much grain as it did in 1967, when observers thought it couldn't produce any more. India is now even the largest *exporter* of rice in the world. Innovation breaks through impossibilities.[6]

Higher crop yields had another advantage: they led to reduced deforestation. If yields had stayed the same from 1960 onward, most of the world's forests would be cut down by now to provide sufficient space to grow enough grains to keep everyone fed. We would have needed additional farmland equivalent to all of the United States, Canada, and China combined. Innovation didn't just help to feed billions of people; it also protected biodiversity and the environment. Yet again, a major challenge was solved not by telling people to make do with less, but by innovating so that there was more for everybody.[7]

The lesson from history is clear. When we innovate and find a cheap, technological solution, we solve major challenges and generate broadly shared benefits. We need to apply that lesson to the problem of climate change.

RIGHT NOW, FOSSIL FUELS are cheap. The global economy relies on them, and there are currently no alternative energy sources fully able to compete. We should be focusing much more on finding and creating those alternatives.

Solar and wind power, so far, are not the answer. Even with huge political support and trillions of dollars in subsidies, solar and wind energy provide just over one percent of our global energy needs. The International Energy Agency estimates that by 2040 and even after another $4 trillion has been spent on additional subsidies, solar and wind power will deliver only less than 5 percent of global energy. They are expensive and inefficient compared to fossil fuels. The cheapness of fossil fuels explains why they meet about 80 percent of our energy needs today, and why they will still be providing 74 percent in 2040, according to the International Energy Agency, even if every promise made by world leaders in the Paris Agreement is delivered.[8]

Reducing our carbon dioxide emissions significantly from fossil fuels will require innovation. The good news is that we have managed to innovate in this area already, almost without trying to. One of the best examples of this in recent times is the evolution of fracking in the United States. Fracking has reduced the price of oil and gas. As we saw in chapter 10, it has lowered the price of heating, allowing poorer people especially to heat their homes better, saving about eleven thousand lives each year. It has also dramatically increased the wealth of the United States—a 2019 study shows that it has increased US GDP by one percent over what it would otherwise have been in 2015, adding $180 billion to the US economy each year.[9]

It has also had real, negative impacts on the environment, especially through air pollution, water pollution, and habitat fragmentation. The biggest study, published in 2019, estimates the total environmental cost of US fracking at $23 billion per year, with air pollution making up three-quarters of that cost. While overall, fracking very likely delivers an overwhelming benefit to the United States, the political conversation for and against it can be seen as a discussion on the distribution of these benefits and costs.[10]

Here, though, we can think about fracking as an example of innovation to cut carbon dioxide emissions. The technology of fracking has been around since at least 1947, but it took public resources to the tune of hundreds of millions of dollars in innovation support from the US Department of Energy and perhaps $10 billion in production tax credits for private entrepreneurs to find ways to innovate procedures to frack gas profitably. The fracking innovation was not intended as climate policy, but simply as a way to make the United States more energy independent and richer. But it also turned out to have a huge climate change benefit, because gas became cheaper than coal. Crucially, gas emits about half the carbon dioxide of coal. Making gas cheaper than coal has shifted a large part of US electricity production from coal to gas. This is the main reason why the United States has seen the largest reduction in carbon dioxide emissions of any nation over the past decade.[11]

One of the most promising innovations that we could achieve in the next decade is to ensure that China makes the same shift. China is by far the world's largest consumer of coal. In fact, in each year since 2011, China has burned more than half of all the coal used worldwide; India (12 percent) and the US (8.4 percent) were in distant second and third places in 2018. If China switched its power production partly to gas, its emission cuts would be massive, dwarfing what we have already seen in the United States.[12]

Sharing with and adapting fracking technology to China and elsewhere is just the start of what is needed. We need to take this experience and look for other areas where the government can support innovation and the development of technologies to ease our transition away from fuels with a high carbon dioxide footprint. To cut emissions significantly, we need to innovate significantly.

UNTIL WE FIND a green energy source that is cheaper than fossil fuels, it will be hard to convince the whole world to fundamentally turn away from fossil fuels. Yet, if we can innovate the price of green energy down below that of fossil fuels, we will be on the pathway to fixing climate change. Everyone, including China, India, and Africa, and not just the United States and Europe, will switch.

It has long been clear that an economically optimal carbon tax will solve only a small part of the climate change problem. That's why I started work with my think tank, the Copenhagen Consensus, back in 2009, to identify other effective solutions to help fix climate change. We worked with twenty-seven of the world's top climate economists and three Nobel laureates to evaluate the costs and benefits of the whole gamut of possible climate change responses. These academics found that green energy research and development is by far the best long-term investment to solve climate change. The experts concluded that globally, we need to spend $100 billion on green energy innovation each year. This would still be much less than what solar and wind energy are costing us in subsidies today, and it would likely substantially bring forward the day when low- or zero-carbon-dioxide energy sources can take over the world.[13]

The economists calculated that for each dollar spent on green energy R&D, we could avoid about $11 of long-term climate change damages. This is a great deal. Moreover, besides helping to find a breakthrough green energy source, this R&D will likely generate many other innovations that can be useful for humanity, such as better batteries for cell phones and cheaper power for space exploration.

Since then, my think tank and many others have campaigned for dramatically ramping up spending on green R&D. The most promising progress came in 2015, when twenty world leaders including President Obama promised to double their country's green energy research and development by 2020. They called their agreement the Mission Innovation.[14]

Unfortunately, those countries have broken that promise. International Energy Agency data shows that in absolute terms, spending has hardly budged, and rich countries are spending less than 3¢ for every $100 of GDP on low-carbon-energy R&D, a percentage that has fundamentally not changed since the Mission Innovation promise was made (see figure 12.1). That adds up to six times less than the amount needed to meet the target of $100 billion per year outlined by the Nobel laureates working with the Copenhagen Consensus.[15]

The frustrating thing is that almost everyone agrees that we should be investing much more in green innovation. It's not a controversial idea. Yet, this spending never seems to be actually allocated. That's because constant scaremongering and green energy industry lobbying lead to scarce resources instead being poured into rolling out more and more of today's inefficient solar panel and wind turbine technology.

Indeed, the green energy industry and its backers often claim that increasing the number of solar panels and wind turbines today is the best way to encourage more innovation, because then the companies can invest in developing better technology. This argument lines the pockets of lobbyists and green energy giants, but it is a hugely inefficient, backward way to go about funding innovation. Globally, private companies spend just $6 billion on renewable energy R&D. As a percentage of global GDP,

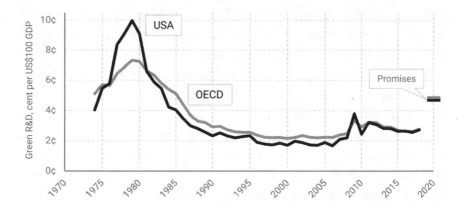

FIGURE 12.1 Green R&D by the United States and OECD from 1974 to 2018 in cents per $100 of GDP. "Promises" signifies what the world leaders promised to spend under Mission Innovation in 2020.[16]

private investment in green energy research has been *declining* since 2012.[17]

To achieve more green energy research and development, we need to spend our money directly on it, rather than relying on indirect industry support and a prayer that it works. Globally, in 2020 taxpayers will pay $141 billion to subsidize inefficient solar and wind energy. This will buy us just $6 billion in actual R&D. Instead, we should spend $100 billion directly on research and development. It will get us $94 billion more for green R&D, and it would still leave us with $41 billion to improve the world in many other ways.[18]

If we want to change the world, we need to commit to doing things differently and coming up with new ideas. Because the old ones aren't working.

WHEN IT COMES to green innovation, what should we focus on?

This isn't an easy question, because no one knows what new technologies will power the world in 2050 or 2100. The prognosis from energy analysts is that thirty years from today we will still be powered mostly by fossil fuels, simply because we know they work and are cheap and reliable. They cause climate change, which is why carbon taxes and other policies will limit their use in the future, but they will remain the main-stay of the world's energy needs. *Unless*, that is, we manage to innovate better, smarter, cheaper, and more effective alternatives.

Predicting future innovations is mostly folly. Before the Chicago World's Fair in 1893, seventy-four of the foremost thinkers in the United States were asked to predict what the world would look like in a hundred years. There were a vast number of memorable misses, such as predicting that the US would see minimal taxation and have no standing army, that everyone would travel everywhere at a hundred miles per hour on electric trains, that war would be abolished, and that we would live to 150 years.[19]

A smaller part of the thinkers got the direction right, even if the tech-nologies were off. E-mail was foreshadowed with the concept of letters sent via pneumatic tubes under the oceans. Air travel would become ubiquitous, although the technology they foresaw was the use of bal-loons guided by wires between cities. Multiple thinkers expected to be able to store energy in small packages that would enable us to deliver the energy of Niagara Falls to manufacturers in Texas.[20]

Many of these thinkers clearly understood the value of innovation. One considered how in a century's time new technology could see houses cooled in summer, just like they were heated by fires in winter. Others re-alized that if electricity became ten times cheaper, it would become uni-versal, driving electrical carriages and easing hard domestic chores like washing clothes, thereby solving the "servant problem." But imagining cheaper electricity or air conditioning doesn't make it happen. Research and development across a wide range of technologies does. Had we just focused on improving pneumatic tubes, we likely would not have devel-oped the internet, nor would a focus on better balloon technology have led to airplanes.[21]

The real lesson is to realize that although we might have a good idea of where we want to go, we have little sense of which technologies will get us there. That is why we shouldn't focus our R&D just on the currently most fashionable and glamorous ideas for the future. Instead we should research *a lot* of ideas.

Research and development is cheap, and a sixfold increase in spend-ing for the world really could allow us to research a wealth of different possible solutions. To show the breadth and opportunity, let me share just three key areas. In each of these areas, breakthroughs could be game changers in the fight against climate change. And in each area, more R&D funding could help stimulate the innovation that could lead to a transformation of our economy.

ENERGY STORAGE IS one area where innovation could make a huge difference to human welfare. If we could store virtually unlimited amounts of wind and solar energy, not only for the seconds or minutes when the wind dies down and the sun is hidden behind a cloud, nor even just for the night or a windless week, but across seasons and even years, storage could deliver energy when we need it rather than when nature deigns to provide it.

When people hear about energy storage, they immediately think of rechargeable batteries like those in electric cars. But these make up less than 1 percent of global energy storage. Almost all storage today, 96 percent, is by the old, well-known technology of water storage. Essentially, this means pumping water up to a reservoir on a hill when there is too much, or "excess," energy, and letting it flow back down through a turbine when the energy is needed. This works well but relies crucially on local geography, and many energy networks are not near suitable sites.[22]

There is a plethora of new technologies that could help to store energy. Compressed air storage pushes air into a salt cavern and then releases it through a generator. A molten salt battery warms up the salt with surplus electricity, and uses a heat exchanger to get electricity back later. Another approach is splitting water with electricity to produce hydrogen, which is later converted back into energy in a fuel cell.

Storage can even be provided with a simple flywheel, essentially a massive rotating drum of steel. The excess energy that might be generated, for example by wind turbines on a windy day, is used to get the wheel spinning. Then a generator captures the energy as electricity again later.

My personal favorite, and this is just because it is a cool idea, is a four-hundred-foot-tall crane standing in the middle of thirty-five-ton concrete cylinders, proposed by the Swiss company Energy Vault. When excess electricity needs to be stored, the crane, controlled by a computer, uses it to stack one concrete cylinder on top of the other, eventually building up a huge tower. When power is scarce, a cable lowering back the very heavy cylinders drives a generator that produces electricity.[23]

The problem with all of these ideas are capacity and cost. If we exclude pumped water storage, all of the other many different solutions in existence today can store just twenty seconds of the world's electricity consumption. Headlines consistently tell stories of how we will see an amazing increase in storage capacity in just a few years. The International Energy Agency estimates that over the next two decades an extra $300 billion will be spent to increase storage capacity forty-one-fold. Yet, this will still increase the available storage to only eleven minutes of the world's electricity use. And storage is still expensive. While utility-scale solar panels can produce power when the sun is shining at as low as 3.2¢ per kilowatt-hour, adding adequate storage more than triples the cost to at least 10.2¢ per kilowatt-hour.[24]

Massively investing in R&D for cheaper and better storage technologies could make a dramatic difference to both capacity and cost in just a few decades. India is already leading the way. It has promised a strong policy push to increase its solar capacity tenfold from 2018 to 2040. (Notice, this won't end the dominance of coal; its use will also double, so it will go from producing 45 percent of India's energy needs in 2018 to 44 percent in 2040.) A lot of extra storage will have to be built to make use of the extra solar power, so India is predicted to become a world leader in storage. Partly as a result, storage prices globally are predicted to halve over the next two decades.[25]

But what if green energy research and development could drive down this storage cost even lower? What if, instead of a 50 percent reduction, extra R&D investment could cut 70 percent off the price of storage? We could then make solar power significantly cheaper than even coal in India. After 2030, India would be able to rely even more on solar and wind power, and stop adding additional coal power (which it isn't forecast to do otherwise). The International Energy Agency estimates that such a low storage price could make India's carbon dioxide emissions peak in 2030, and by 2040 this alone would shave almost one percent off global emissions.[26]

Imagine how much more attractive solar power would look to the world's other developing countries, if storage was effective and afford able. Innovation in storage could help drive down global emissions, make power cheaper, and unlock more of the potential of fossil fuel al ternatives.

THE SECOND OBVIOUS AREA for R&D investment is nuclear en ergy. Nuclear energy doesn't emit carbon dioxide. Perhaps surprisingly, nuclear energy is also very safe. Under normal operating conditions, it emits less radioactivity than coal (yes, there is actually radioactivity in mined coal that is released when it is burned). Although we think of the terrible disasters of Fukushima and Chernobyl, nuclear energy has one of the lowest death risks of any form of energy; in fact, it kills about two thousand times fewer people than coal power, because of coal's massive pollution.[27]

The reason why nuclear energy isn't a silver bullet in its current form is that new nuclear power plants in developed countries are *much* more expensive than continuing to rely on fossil fuels. Finland's newest nu clear power unit was supposed to open in 2010 and cost $3.5 billion. It may open in 2021 at triple that amount. France's Flamanville nuclear power plant was supposed to open in 2012. It may open in 2022, again at triple the cost.[28]

A 2017 survey shows that in the 1950s and 1960s, when nuclear power was first developed, plants got less and less expensive as more were built, as experts learned how to do so more cheaply and efficiently. But in the 1970s and onward, nuclear power plants became ever more ex pensive to build. Why? In France and the United States, instead of fo cusing on a single, standardized unit, designs began to change, both because of an interest in improving them and because of increasing regulations. One 2017 study suggests that had we been able to stick to one or just a few designs for nuclear power plants, ensuring that costs would continue to fall, nuclear energy today could cost-effectively re place all coal and most gas power, and reduce global carbon emissions by one-fourth.[29]

Of course, we can't run the clock backward, but we can make smarter decisions moving forward. That means investing in innovation to make the next generation of nuclear power plants much cheaper and safer.

Philanthropist Bill Gates is investing in TerraPower, a company developing new reactors that promise to use the waste of other reactors to generate energy for decades without additional fuel. China is demonstrating a "pebble bed" design intended to be very safe. Others are suggesting modular designs that can be built from standardized reactor parts in a factory and put together like Legos at the construction site.[30]

One 2019 study shows that these new ideas could cut the cost per kilowatt-hour of nuclear-generated electricity by almost two-thirds of today's cost, and the most optimistic cost estimates would see nuclear power become cheaper than even the cheapest gas power. Of course, these are estimates of cost reductions from design improvements. If we manage to uncover a revolutionary breakthrough, the cost reductions could be even greater.[31]

Nuclear innovation could help us find a way to power the world more cheaply and safely than today, without carbon dioxide emissions.

A THIRD AREA where additional research could make a huge impact is air capture. This is simply machinery on the ground that sucks out carbon dioxide from the air and stores it safely. If this could work, we could achieve some, or even all, of our carbon emission reductions without cutting fossil fuel use. In an extreme (and unrealistic) case, we could maintain our entire global fossil fuel economy, and simply suck out the problematic emissions.[32]

This is often what offsets do today: they represent a pledge that trees will be planted to soak up carbon dioxide emissions. Essentially, trees are the low-tech version of air capture.

Why not just plant more trees? Unfortunately, there is not enough land on the planet to actually capture all the excess carbon dioxide, and moreover we need land for agriculture. Air capture could be a much more efficient solution, but there is still a long way to go. In 2007, entrepreneur Richard Branson and climate change campaigner former vice president

Al Gore proposed the Virgin Earth Challenge, a $25 million prize to be awarded to the first team to create a scalable air capture option. The prize givers have yet to be presented with a solution that satisfies all the conditions to win.[33]

In 2011, an influential panel of the American Physical Society found that air capture would cost $600 or more per ton of carbon dioxide. That means it would cost $5.33 to suck up the emissions from just one gallon of gasoline. Obviously, at that price the technology is not going to get very far. Since then, though, researchers have explored many ideas. A 2017 overview shows a huge variance in projected costs, with some start-up companies suggesting they can achieve air capture with almost zero costs. We should take these claims with a grain of salt but promisingly, a 2018 peer-reviewed study showed that a viable project could possibly cost less than $100 per ton, or 83¢ per gallon of gasoline.[34]

Spending more money to innovate to drive down the cost of air capture could be tremendously advantageous. Right now, if we were to offset all of humanity's emissions at $100 per ton, it would cost $5.5 trillion every year, or more than 6 percent of global GDP. Imagine if innovation could drive down the cost of air capture to, say, $5 a ton by midcentury. Avoiding the emissions from a gallon of gasoline would then cost just 5¢, which it's plausible to imagine most people being willing to pay. If we can achieve that low cost, we could eliminate practically *all* of the world's emissions from the atmosphere for just 0.2 percent of global GDP.[35]

Air capture today is so expensive that it cannot yet be a serious part of our response to climate change. But innovation could help lower its price to a point where it could be an important tool in our response to the problem.

THERE ARE MANY, many more breakthrough technologies that could possibly play an important role in solving climate change. Craig Venter, the geneticist who led the first draft sequencing of the human genome, argues for research into algae that can be grown on the ocean surface to produce oil. Because the algae convert sunlight and carbon dioxide into oil, burning it will be carbon dioxide neutral. This technology is not yet efficient, but more research and development could potentially make it so. Fusion power, mimicking the sun, could provide all the energy humanity needs, with no carbon dioxide whatsoever. But while fusion power is nowhere near commercially viable, again, more R&D could perhaps make it so.[36]

Indeed, every single one of the ideas we have discussed could fix a large part or even *all* of climate change. They are not yet ready for global rollout because they remain too expensive, but research and development could push their price down.

The good thing about R&D is that it is relatively cheap. A hundred billion dollars could fund innovation across a broad spectrum of potential technologies. We should expect most of these ideas to fail. Many will see some progress, but not enough to become economically competitive. Air capture's cost might fall to $70 per ton, say, but remain too expensive to be used everywhere across the planet. The cost of nuclear power might be driven down to 6¢ per kilowatt-hour, but still not be cheap enough to replace fossil fuels. The Swiss stacking experiment to store energy could, like any other individual idea, fail to become anywhere close to cost effective.

But this is okay. When we invest in innovation, most ideas fail. We don't need a lot of technologies to power the world. We don't need many ways to cut carbon dioxide emissions to zero. We just need one (or more realistically, a few).

Imagine if innovation could lead to such a massive breakthrough in air capture technology that it is priced at $5 per ton. Fixing global warming would cost less than what the EU pays for its climate policies today. Of course, there would still be squabbling among nations about who should pay what. But we would fix climate change.[37]

Imagine if innovation could revolutionize the cost of nuclear power down to 1¢ per kilowatt-hour. Not only would the entire world switch to this much cheaper alternative and essentially solve global warming, but also it would be an incredible boon for humanity, providing more energy, with cleaner air, at almost zero cost.

What should the model for innovation look like? Over the past half century, governments have developed an approach to drive innovation through initial, public "blue sky" research in universities and research labs. Some of these innovations have, in partnership with private businesses, been developed to the point where they can then be profitably introduced by private companies.

We should recognize that different countries have different strengths in R&D areas, so individual countries could focus on different topics. The smartest way forward would probably be the establishment of national, blue-ribbon panels to identify interesting energy research problems that need solving. Funding could be managed through national organizations such as, in the United States, the National Science Foundation.

Individuals and groups of researchers could think of ideas that might solve part of one of the research problems outlined by the blue-ribbon panels—ideas from oil-producing algae, to tidal power, to novel carbon dioxide capture systems such as enhanced weathering rocks, to new nuclear reactor designs, and everything in between. These researchers would apply for funding to solve specific problems. Their peers would decide which applicants would be more likely to drive research and understanding within the particular field, so as to help improve the chances that in the long run, ideas could turn into real progress for clean energy.

But none of this can happen unless we focus on this one thing: we need to ensure that R&D investments increase dramatically. In just a year or two, the United States and other nations need to at least double their annual investments in order to meet the level promised in the Mission Innovation. Over the next five to ten years, R&D spending globally should increase a further three times to get close to the $100 billion annually identified by the Nobel laureates. With a sixfold increase in our budget, we could help fund so much more. We could dramatically improve humanity's chances of innovating down the price of future solutions to help address or entirely fix climate change.

Innovation is not the only solution we should pursue to tackle global warming, but it is one of the most promising. Finding ways to produce green energy cheaper than fossil fuels, or finding very cheap ways to avoid the environmental impact of fossil fuels, would transform the fight against climate change. It would eliminate the tension that exists right now between the twin goals of growth and reducing emissions. Through innovation, we could solve the problem of fossil fuels in the old-fashioned and proven way—by making the alternatives cheaper and better.

ADAPTATION:
SIMPLE BUT EFFECTIVE

A WELL-CONCEIVED CARBON TAX can help us avoid the worst climate change damage, and a large investment in innovation can bring forward the day when the global economy ends its reliance on fossil fuels. But even with both of these policies, the temperature will continue to rise. Some of future warming comes from past carbon emissions, which we have no control over today. And neither carbon taxes nor innovation will entirely eradicate future emissions, at least not in the short to middle term.

This is why we need to adapt to a warmer planet over the coming decades. Fortunately, humanity has remarkable adaptive capacity. There are people living in the icy extremes of Siberia and northern Canada, in the burning hot Sahel desert and Australian Outback, in the dry Atacama desert plateau of South America, and in the rain-soaked Meghalaya state in India. Not only do people withstand significant variations in temperature and rainfall, but falling per capita death rates from natural disasters show that we have more resilience today than ever before.[1]

At its simplest, adaptation means that people react sensibly to challenges—in our case, to a changing climate. As it gets warmer, more people will adapt by turning on their air conditioners (and fewer will use their heaters). If they don't yet have an air conditioner, more people will buy one (and more people will be able to buy one as global prosperity increases). Similarly, tourists will adapt to a warming world by changing their travel destinations. Warm places like Sri Lanka will host fewer tourists. On the other hand, more visitors will choose Finland and Canada for their next holiday, while fewer Finns and Canadians will travel abroad.[2]

Cutting emissions has significant costs, but helps everyone in the world a little, albeit a half century down the line. In contrast, adaptation often has immediate and very localized benefits. Indeed, most adaptation happens naturally and needs little public policy focus or investment.[3]

Businesses often don't need to be forced to invest in adaptation, because it makes sound financial sense. There are plenty of examples of this already. In the rich world the chemical giant BASF has installed additional water pumps in the river Rhine, so that even if climate change lowers water levels, there will be enough water for production. Unilever works with tomato suppliers to encourage them to install drip irrigation so they can grow crops even in droughts.[4]

In poorer countries, farmers also adapt to a changing climate. In South America, a study shows that farmers already tailor what they grow to the climate: farmers grow fruit and vegetables in warmer locations, and wheat and potatoes in cooler locations. Where it is wetter, they grow rice, fruits, and potatoes, and in drier locations they grow corn and wheat. As climate change drives up temperatures, farmers will adapt by switching to more fruits and vegetables, and depending on whether it gets wetter or drier (the models are still not sure), they will add potatoes or squash.[5]

But not all needed adaptation will happen without specific government actions. Clearly, governments should make sure that public policies do not hinder private adaptation, for instance by levying high taxes on air conditioners or the electricity required to run them. But governments should do more: they should implement policies that make adaptation easier. Across the world, agricultural adaptation is easier if you are better educated, if you are better off (for instance, if you have a tractor), and if you have better access to agricultural information. So governments would do well to make sure there is better access to education, agricultural information—and tractors.[6]

In Ethiopia, a 2011 study found that farmers with access to credit can better adapt to a changing climate, and they end up with higher food productivity. This is not surprising; if you can get extra resources to over-come a challenge, you will be much more likely to do well. So, govern-ments should make sure that credit opportunities are as widely available as possible. This doesn't mean subsidizing the actual loans, but rather ensuring a well-functioning legal and institutional framework that will make it easier for individuals to access the necessary funds to adapt.[7]

In some areas, adaptation relies on public policy. It is not realistic to expect individuals to adapt their houses to rising sea levels or their lives to the threat of catastrophes. Governments need to step up with flood defenses and early-warning systems. And while air conditioning can help during heat waves, the right infrastructure can make whole cit-ies cool. Moreover, while many people can adapt on their own, the most vulnerable often can't. Public policy can be especially useful in helping elderly and marginalized people when heat waves hit.

Pursuing policies that enable adaptation seems like common sense. Yet weirdly, for a long time in climate change policy discussion, it was considered bad form to even mention adaptation. Climate change cam-paigners have tended to view the idea of adaptation as distracting atten-tion from cutting carbon dioxide emissions. Perhaps they also believe that acknowledging the need for adaptation is an admission of defeat in the battle against climate change.[8]

To the contrary, if we are to deal effectively with climate change, we need to put adaptation at the heart of our policy response, right along-side a carbon tax and innovation.

RISING SEA LEVELS get a huge amount of attention in the media, and they are often portrayed as uncharted territory for humanity. In fact, sea levels have risen about a foot over the past 150 years. Around the world, when you ask anyone what important events happened over that cen-tury and a half, they will talk about wars, medical breakthroughs that saved lives, perhaps the moon landing, but they won't tell you that rising sea levels were a big deal. Why? Because we adapted to them by protect-ing our coastlines.[9]

One of the clearest results of studies of adaptation around the world is that coastal protection for populations and valuable land is a great investment. As we saw in chapter 1, coastal protection costing tens of billions of dollars can avoid tens of trillions of dollars in flood damages. That is why a recent 2018 overview shows that the cost of sea level adaptation almost everywhere will be much lower than that of not adapting. The study shows that even if the seas were to rise an improbable six feet seven inches by 2100—vastly more than the UN panel of climate scientists expects—it will be economically advantageous to protect at least 90 percent of the global coastal floodplain population, along with 96 percent of all assets there.[10]

For more than half of the population of coastal floodplains throughout the world, each dollar spent on protection will avoid more than $100 of damage. The total cost of coastal protection *and* of all remaining flood damage through the rest of the century, even in the absolutely worst-case scenario, will cost the United States just 0.037 percent of its GDP, and possibly five times less.[11]

"Coastal defense" in many cases will mean dikes (a long wall or embankment built to prevent flooding), but softer approaches like artificial nourishment (meaning adding sand to beaches) can be even more effective in dealing with the impacts of rising sea levels and storm surges. In a 2019 overview of nineteen studies, dikes on average reduce damages by $40 for each dollar spent, whereas artificial nourishment can avoid $111 of damages for every dollar spent.[12]

Other natural defenses like the restoration of mangroves can help, too. As well as providing a buffer against tidal storm surges, mangroves provide critical habitat to sustain local fisheries. Planting (or reestablishing) mangrove forests, as is being done in Indonesia and elsewhere, is several times cheaper than building flood protection infrastructure. The benefits of mangrove preservation and restoration are worth up to ten times the cost, including not just avoided losses from coastal flooding, but also the benefits associated with fisheries, forestry, and recreation.[13]

People talk about rising sea levels in apocalyptic terms, but the truth is that there are proven and cost-effective ways we can defend coastlines to ensure that more people and possessions are protected.

AS GLOBAL WARMING leads to an increase in heavy precipitation, we may well see more river flooding and flash floods. Adaptation can help, but it needs to be coordinated. If everyone tries to dam the river, it becomes an unpleasant game of "who has the lowest levee."

Holland has shown the way by giving "room for the river," a program that allows some floodplains to be flooded so the flood waters don't destroy cities, and that deepens and widens rivers so there is room for more water, making them flood less. The floodplains can even be turned into parks. In the Netherlands city of Nijmegen, close to Germany, a new river park and riverfront development improved the local quality of life, while at the same time expanding floodplains.[14]

Across the United States, existing river flood adaptation definitely works. If there were no flood defenses, nearly twenty-two million people would experience flooding each year, but with defenses the actual number is much lower at half a million people. The National Institute of Building Sciences finds that each dollar spent adapting to floods in the US provides benefits worth $6. For the EU, it is estimated that future benefits of adaptation to more floods will outweigh the costs seven to one.[15]

Houston, Texas, is a good example of a city undertaking actions to reduce its vulnerability to flooding. It has had far more than its fair share of floods and climate-related catastrophes. It is in fact the US city most severely affected by floods. Why? Partly it's because it is very flat and built on a swamp. But the city was also very poorly planned. The population has grown rapidly, and infrastructure—roads, parks, and sewers—didn't keep pace.[16]

Between 2000 and 2018, Houston's metropolitan population jumped from around 4.5 million to nearly 6.3 million. The city today sprawls over 580 square miles. To claim that space, the city paved over a vast amount of wetlands and prairie land. These are beneficial because they soak up huge amounts of rainfall, so they are crucial to avoiding floods. When they are paved over, heavy rain means that vast expanses of concrete create runoff. Since the 1980s, rainfall has increased 26 percent in one Houston watershed, but runoff has increased by 204 percent. One 2018 study shows that runoff exposed an additional thirty-five hundred households in the Sims Bayou watershed in the south of the city to flooding.[17]

In 2019, Houston decided to use $1.7 billion from its appropriately named Rainy Day Fund savings on flood protection. Much of this is being spent on big infrastructure: dams and levees, which are important. A 2019 study shows that building a coastal barrier to protect Houston could cost $400 billion over the century, but avoid damages around twice that much.[18]

But there are also many other, simple approaches that can do a lot of good. One proposal suggests buying out and demolishing houses on the floodplains to re-create more absorbent, green spaces. In one area in North Houston, the city is already planning to buy out two low-income apartment complexes and covert the area into a flood basin and park. But the properties being bought out are only a small fraction of those in flood-prone areas. Houston could also build more resilient infrastructure into those neighborhoods by implementing "pervious" pavements that can soak up rain, and by creating better drainage at the sides of roads.[19]

Across the world, there are many sensible adaptations to avoid flood damage. Planting trees to soak up extra water is estimated to avoid two dollars of damages for every one dollar spent. Ensuring that buildings have gutters that collect and reuse rainwater means the sewers will be less overloaded, with slightly more benefits than costs. Encouraging homeowners to make basic changes to their homes and grounds, such as planting trees and making use of rainwater, turns out to be both cheap and effective. Of course, cities like Houston can also choose to clamp down on future building to ensure the few wetlands and prairies that are left can still play a role in reducing floods.[20]

In short, to reduce flood damages, local and national governments have an array of cost-effective tools at their disposal.

BEYOND FLOODING, ADAPTATION can protect us from other natural disasters, including hurricanes. Bangladesh is more vulnerable to hurricanes than most countries, but it has reduced its vulnerability with effective investment in adaptation. The country is the size of Illinois, but has thirteen times the population, much of it extremely poor. It is located at the upper end of the Bay of Bengal, which can funnel hurricanes into the country and amplify flooding. Early on November 12, 1970, the Bhola hurricane struck the area as a category 3 storm and became the world's deadliest hurricane. Radio broadcasts only started to warn citizens of danger late on the afternoon before, and most people disregarded these messages because of mistrust, fear of theft if they left their homes, and because there were few shelters available to go to in any case. Although some 90 percent of the population heard the warnings, just 1 percent fled.[21]

The hurricane was ferocious. Its winds reached 150 miles per hour, striking at high tide and creating a tidal wave twenty feet high. At least 250,000 people were swept away—and quite likely twice that number. The disaster, along with the government neglect that exacerbated it, was a major factor in Bangladesh (then known as East Pakistan) breaking away from Pakistan the following year.

The new nation created the Cyclone Preparedness Program, and over time invested in early-warning systems and raising civic awareness, as well as in recovery services and the creation of many new storm shelters and reinforced buildings.

In 1991, at which point Bangladesh had still built only three hundred shelters, it was hit by a much stronger, category 5 hurricane. The death toll was half that of 1970 because the Cyclone Preparedness Program worked. Since 1991, a strong focus on adaptation and the construction of a further thirty-five hundred shelters have seen the death rate cut by more than a hundredfold. In the last three decades of the last century, about fifteen thousand people died in Bangladesh each year because of hurricanes. In the 2010s, thanks to widespread adaptation, the average number is just twelve dead.[22]

Overall, it has been estimated that modest investment of around $1 billion a year to improve public awareness, early-warning systems, and disaster response in developing countries could deliver total benefits ranging from $4 billion to $36 billion. Indeed, studies show that across a wide range of climate impacts, adaptation through greater public awareness of dangers and early-warning systems can be an excellent investment. For river flooding, warning systems can reduce damage costs by four times the cost of the warning. For heavy precipitation, preparing communities through training and emergency management can return $30 on every dollar spent.[23]

As we saw in chapter 4, climate change is likely to mean over time that hurricanes are less numerous but more ferocious. Adaptive measures are necessary to ensure more people are protected, and they are well within our power—and our budget—to implement.

WE CAN DO much more, too, to reduce the risks of forest fires through adaptation, especially in places like California, where the Camp Fire burned down the entire town of Paradise in 2018.

One of the best ways to prevent fires from destroying lives and property is to make sure that people don't put themselves in harm's way by building homes in high-risk areas. Without so many homes being constructed in vulnerable areas at the edge of the forest, fires would still happen, but without the human devastation.

Building codes matter too: California has strict codes for new buildings, but older homes with more flammable roofs are quicker to catch fire, endangering the houses around them. Indeed, the state's so-called wildland-urban interface code highlights how to build safer houses by ensuring better placement, more fire-resistant building materials, sprinklers, and less vegetation. If this code was adopted throughout the United States, it is estimated that each dollar spent meeting code would avoid four dollars in fire damage, and in Florida and the West avoid more than six dollars.[24]

To see the difference that adaptation makes, we need only look to the city of Montecito, 450 miles south of Paradise. The city suffered severe fires in the 1960s; in response, Montecito adopted an approach called "adaptive resilience." This strategy included creating spaces around homes without any wood that could catch on fire. In some places, cactuses were used to replace fire-prone trees. Vegetation on the sides of roads that could create a terrifying "tunnel of fire" for people escaping were removed. And existing homes were "hardened" by making sure they used fire-resistant building materials. In addition, firefighters themselves set carefully prescribed fires to reduce the amount of wood across the countryside. And they implemented detailed fire-planning and response outreach programs.[25]

In 2017, all these actions were put to the test. The Thomas Fire, which had already been burning for nearly two weeks, turned south, and sixty-five-mile-per-hour gusts of wind began blowing the fire directly toward Montecito. There was an expectation that hundreds of houses would be burned to the ground; in fact, only seven were lost. A fire expert said of the efforts to protect Montecito that "it's the difference between living in a matchbox and a place that's more resistant."[26]

Adopting the adaptive measures undertaken in Montecito and rolling out stricter codes for homes in harm's way is just common sense.

HEAT WAVES INVARIABLY make the news and are often invoked as evidence of the coming climate change apocalypse. Temperatures are increasing—there's no question—and extreme heat will become a greater problem in the decades and centuries to come. But as with so much else, we have tools already in place to mitigate much of the impact.

Adaptation should start with cities, because rising temperatures will have the biggest human health impact in urban environments. Cities are increasingly where we live, already accounting for more than half the world's population, and by the end of the century the figure will be 80 to 90 percent. Moreover, cities are generally much hotter than surrounding countryside: they are filled with masses of nonreflective black surfaces that absorb the sun's rays and lack green spaces and water features.[27]

This so-called urban heat effect can be drastic. For example, during the summer, Las Vegas is 7.3°F hotter than its surrounding countryside, with summer nights an astounding 10.3°F warmer. All the asphalt and vast buildings have made the city's night temperatures increase much faster than the surrounding countryside, a rise of almost 1°F each decade since 1970.[28]

But there is an amazingly simple adaptation that can make Las Vegas and other cities cooler. Heat is caused by black roofs and black roads, so we should make roofs and roads lighter in color. Heat is exacerbated by an absence of parks and water features, so we should create more greenery and oases.

Black roofs are a human health risk. In the aftermath of the 1995 Chicago killer heat wave, analysis showed that people living on the top floor of a building with a black roof were more likely to die. But lessons have been learned. New York is working with nonprofit organizations and building owners to apply white reflective surfaces to rooftops. It has already painted more than five million square feet of roofs with a light, reflective coating (but that is only a tiny fraction of the city's total rooftop space). Los Angeles has started painting its dark asphalt with a cool, gray coating that can lower the asphalt's temperature by 10°F. The city calls these "cool pavements." Theoretical models show that making roads and roofs lighter in color can reduce summer temperatures in California by 2.6°F and in New York by 3.2°F.[29]

Planting trees and expanding green spaces and water features not only make for a far more pleasant city for inhabitants, but also drastically cool the highest temperatures. In London, the area around the River Thames and in urban parks is on average 1.1°F cooler than neighboring built-up areas.[30]

The most significant impact of these approaches will be evident during actual heat waves. If we increase parks and water features significantly, models show, high temperatures three days into a hot-weather spell can be decreased by as much as 14°F, providing an oasis of coolness. And analysis from London indicates that painting asphalt and black buildings white, and thus changing the entire city's reflectivity, could decrease the temperature three days into a heat wave by a whopping 18°F.[31]

Adaptation is very effective at cooling cities. One 2017 study shows that large-scale adoption of cool roofs and pavements across the world will cost about $1.2 trillion over the century, but it will prevent climate damages worth almost fifteen times as much. Too often, however, the media treats cool roofs and light-colored pavements as quirky oddities rather than serious policies. The media pays far more attention to "green roofs" with gardens, water, mosses and plants, presumably because these urban oases seem much more charming. Unfortunately, green roofs are a poor deal in regards to the climate: their benefits are no greater than simple cool roofs, but they triple the cost.[32]

There are many other ways of adapting to rising temperatures. Information is a crucial one. Better weather forecasting can help identify the risks, and campaigns can encourage citizens to take simple measures like using fans, drinking plenty of water, and wearing hats.

While some people will be able to take measures to adapt to heat waves by themselves, public measures are also essential. Those with swimming pools can use them to beat the heat, but for everyone else, keeping public swimming pools open for longer hours during a heat wave can help save lives.[33]

In the United States, local authorities already work with national agencies to determine when a heat wave is approaching and issue heat alerts along with educational information so that people can plan and keep themselves safe. In Chicago, such actions led to deaths dropping from 700 to 100 in three years, while a Philadelphia heat wave plan is estimated to have saved 117 lives over three years. France has taken similar actions and managed to reduce deaths during heat waves by 90 percent.[34]

These adaptive measures are working not just in rich countries but also in poor countries. After a 2010 heat wave killed more than thirteen hundred people, the Indian city of Ahmedabad enacted a Heat Action Plan that included training health care staff, distributing water, and painting roofs with white reflective paint to make homes up to 9°F cooler. A similar heat wave in 2015 claimed fewer than twenty lives. Unsurprisingly, the Heat Action Plan has been copied by other cities across India.[35]

Large numbers of deaths from heat waves are entirely avoidable if we take sensible adaptive measures.

ALONG WITH A carbon tax and green innovation, adaptation is crucial if we are to tackle climate change. As we've seen, much of this will happen by itself, simply because people naturally adapt, including in the simplest ways from carrying an umbrella to avoid getting soaked to turning on an air conditioner to avoid getting heatstroke. But there is still a lot of publicly funded adaptation that can reduce climate damages, including better coastal and river protection, early-warning systems for disasters, fire-proofing communities, and taking steps to cool cities. Most of these are very cost-effective investments with considerable returns to society.

Most of the alarmism about climate change tends to ignore our ability to adapt. Remember the claims in chapter 1 that coastal flooding in the United States could cost more than our entire current GDP? Such scaremongering relies on the assumption that we will not adapt. But clearly, adaptation will happen, and it will not be through some global treaty, but through local and national decisions. That's happening already.

Take the island of Manhattan. When Hurricane Sandy hit New York City in 2012, lower Manhattan suffered expensive flooding damage because basic solutions like storm covers for the subway system were missing. In 2019, Mayor Bill de Blasio set out a comprehensive plan to protect lower Manhattan against future flooding damage by augmenting berms and storm barriers, elevating parks, creating a coastal extension, and more. This is on top of other initiatives that include building a five-mile seawall around Staten Island and sand dunes around the Rockaways.[36]

Clearly, poorer people have fewer resources to spend on adaptation. When a hurricane hits poor shanty towns, many people die. When a hurricane hits rich Florida, it might have a severe economic impact, but the human devastation will be far less, because people can afford much more adaptation. Similarly, we saw that Montecito and Paradise experienced very different fates from their fires. Yes, the vegetation that surrounds them differs, but there is another difference that played a crucial role in the final outcome: wealth. Paradise is a working-class community, whereas Montecito is one of America's wealthiest enclaves. To maximize everyone's chances of adapting and reducing their vulnerability, helping to achieve more prosperity is a good place to start.[37]

Perhaps the most remarkable fact about adaptation is that most of its benefits can be realized fairly cheaply within just a few days or a few years. Compare this speed to the delayed impact of worldwide carbon taxes. Adaptive actions can typically deliver much more protection much faster and at a lower cost than any realistic carbon-reduction climate policy.

We should invest far more in planning and infrastructure to provide protection from natural disasters, rising sea levels, and changes in temperatures. We must do all of this with a clear understanding that adaptation is an effective and necessary climate policy.

GEOENGINEERING:
A BACKUP PLAN

CARBON TAXES, INNOVATION, and adaptation can reduce carbon dioxide emissions, bring forward the day that we transition away from dependence on fossil fuels, and reduce our vulnerability to climate change. However, our track record in delivering effective climate policies is not stellar—while we have definitely managed to adapt, carbon taxes and innovation have still not been pushed sufficiently. But there is another approach that could dramatically reduce temperature at a very low cost, and in as little as weeks. It is called "geoengineering," and essentially means deliberately adjusting the planet's temperature controls.

With this solution more than any other, we enter uncharted territory. Humanity has never purposefully made planetwide efforts to change the climate. Many geoengineering technologies sound like science fiction. Understandably, the entire field of study provokes fear.

It is not a policy we should be *implementing* right now. But geoengineering is a partial solution to climate change that is worth *researching*. We should think of geoengineering as a backup plan that we could turn to, if we don't manage to get everyone to do everything needed on carbon taxes, innovation, and adaptation. And because it works so quickly, it could play a role if we found that we needed fast action to avoid a looming catastrophe.

OVER A LONG WEEKEND in June 1991, Mount Pinatubo in the Philippines erupted. It was by far the largest volcanic eruption to affect a densely populated area in the twentieth century. Producing avalanches of hot ash and gas, the eruption killed hundreds, damaged thousands of homes, and displaced hundreds of thousands of people.

As well as causing devastation, the volcano also affected the climate. The eruption injected enough sulfur dioxide into the stratosphere to temporarily reduce the amount of sunlight reaching the earth's surface by about 2.5 percent; as a result, temperatures around the globe dropped by an average of one full degree Fahrenheit over the following eighteen months.[1]

As concerns about global warming increased, researchers began to investigate whether or not they could mimic a volcano's effect on the climate, without the carnage. It can be done with something called "stratospheric aerosol injection," which involves spraying tiny particles such as sulfur dioxide into the upper layer of the atmosphere to act as a thin reflective barrier against incoming sunlight. Scientists have come up with a number of proposed delivery mechanisms to get the sulfur dioxide where it needs to go—from using powerful artillery guns, to long pipes suspended by high-altitude balloons, to aircraft that disperse particles as they fly.[2]

The idea of pumping sulfur dioxide into the upper layers of the atmosphere understandably worries many people. Some fear that darkening the sky in one hemisphere would have huge, unpredictable effects on tropical climate patterns, and perhaps lead to more drought in the Sahel region. Others worry that it will hamper photosynthesis. But scientists are investigating other geoengineering options, too.[3]

One of the cheapest and most effective approaches is called "marine cloud brightening." Waves breaking on the ocean create airborne sea salt particles, and the clouds above oceans consist mostly of tiny water droplets that have condensed around these particles. The idea is that if you can increase the number of sea salt particles in the air over the oceans, the resulting clouds will end up with more tiny water droplets. Fewer larger droplets mean darker clouds (as we know from anytime it is about to rain), while many smaller droplets make clouds whiter. If we can make more ocean clouds whiter, they will reflect more solar energy back into space, thus cooling the planet.[4]

Marine cloud brightening amplifies a natural process and would not lead to permanent atmospheric changes, since switching off the entire process would return the world to its previous state within a few days. It could therefore be used only when needed.

Together with John Latham of the National Center for Atmospheric Research, the University of Edinburgh's Stephen Salter has designed a fleet of remote-controlled, wind-powered catamarans that could mimic the ocean's natural wave action and put more sea salt particles into the atmosphere. This fleet would disperse seawater mist about a hundred feet into the air, introducing more sea salt particulates and helping clouds become slightly brighter—just enough to keep temperatures down.[5]

THE DELIBERATE MODIFICATION of the climate to suit human needs has long been regarded as anathema, or at least the height of hubris, by most environmentalists. Climate is one of the most complex systems imaginable, and we are a long way from fully understanding how it works. Who is to say that in our well-intentioned efforts, we won't make things worse—maybe even much worse?

Obviously there is a lot that we do not yet know about this technology and how it would work on a planetary scale. But there are three key reasons why we should start researching geoengineering technologies.

The first argument for researching geoengineering is that it is the only known approach that allows us to make dramatic cuts in global temperature at low cost. Research for Copenhagen Consensus shows that just $9 billion spent building nineteen hundred seawater-spraying boats could prevent *all* of the temperature increase projected in this century. This is a tantalizing possibility when we consider the $60 trillion in damages in the twenty-first century that would come from unmitigated global warming.[6]

The second argument follows straight from the first. If changing the planet's temperature turns out to be relatively cheap and easy, there is a risk that a single nation, a rogue billionaire, or even a highly energized nongovernmental organization could deploy the technology on their own. Thus, it is important that scientists seriously investigate its potential impacts now and share this information, to make sure we all know more about geoengineering's potential negative impacts. If there are severe, negative consequences it would make rogue efforts to pursue geoengineering less likely.[7]

The third key argument for exploring this technology is that it would allow us to change the global average temperature very quickly. Any standard fossil-fuel-cutting policy will take decades to implement and half a century to have any noticeable climate impact. Instead, just like Mount Pinatubo, geoengineering can literally reduce temperatures in a matter of weeks.

There is a great deal we don't know about climate change, including whether or not it happens in a linear fashion. Proponents of drastic carbon cuts often make the case that there could be tipping points that we barely understand, beyond which the planet could be doomed. What if they're right? What if we learn at some point that we really are five years from disaster? None of the other weapons in our arsenal work on that time scale. Even radical cuts of emissions would take decades to achieve significant results. Geoengineering is the only way to halt warming quickly. Potential pitfalls are clear, but if we were faced with a genuine catastrophe, we would certainly want this option available.[8]

THERE IS WIDESPREAD opposition to geoengineering from those you may think would be most in favor: people who have devoted their careers to tackling climate change. A 2019 survey of more than seven hundred climate change scientists and negotiators involved in international climate policy found that most believe we should not study geoengineering. (Interestingly though, when those surveyed believe climate change will be bad for their *own* country, they are much more likely to support geoengineering.)[9]

The key arguments made against geoengineering are that the technology won't work, that it will work but will have bad impacts, and that it will distract us from cutting carbon dioxide emissions.[10]

The first argument is simply an empirical question. Maybe geoengineering works, maybe it doesn't. To find out, we need scientists to undertake research. Will geoengineering cause harm? Maybe, which is all the more reason we should research it. Of course, there are many things that could go wrong. We could change complex precipitation patterns, for example, meaning that rainfall increases where we don't want it, and decreases where it's needed. If we reduce sunlight as stratospheric aerosol injection would, plants would grow less quickly, so that could lead to lower agricultural yields.[11]

However, the research done so far is actually encouraging. One team of researchers looked at two possible worlds by the end of the century. In one of these we experience high temperatures because of global warming. The other is identical except the world has started using stratospheric aerosol injection, following the concept of Mount Pinatubo. In the vast majority of places people live, the second scenario is better than the first: only a few places, less than 0.4 percent of the planet's inhabited surfaces, have more extreme weather. Most people experience less extreme weather, including lower temperatures and lower extreme temperatures, along with less extreme precipitation and less risk of flooding or drought.[12]

But the third argument, that somehow geoengineering is bad because it undermines the "real" solution to climate change, is deeply flawed. The same reasoning was once used to dismiss any discussion of adaptation. Campaigners who argue this point believe that there is only one "right" way to fix climate change, namely, cutting carbon dioxide emissions. Surely, we should be open to using whatever method will be most effective.[13]

Moreover, it is unreasonable to argue that using geoengineering will just give us an excuse to keep emitting carbon. Should we ban coronary bypass surgery or cholesterol-lowering drugs because they let people get away with bad behavior like eating too many french fries? In my experience, nobody eats french fries just because they know that if it comes to it, they have an option down the line to get a bypass.

Even in the face of opposition from many climate change campaigners, it is promising to see increasing support for geoengineering research from various quarters: under President Obama, the US government office that oversees federally funded climate studies actually recommended geoengineering research, marking the first time that scientists in the executive branch ever formally called for it. This also had the backing of

President Obama's science advisor, John Holdren. And in the wake of the Paris climate deal, eleven top climate scientists declared that the Paris Agreement was far too weak to prevent the effects of climate change, and announced that they supported geoengineering research.[14]

IN A COST-BENEFIT STUDY undertaken for Copenhagen Consensus, researchers estimated the benefit of research on geoengineering. They found that every dollar spent on research in this area could lead to a total benefit worth $2,000. Yet astonishingly, given the potential, very little government money has been put into researching the technology. One of the very few organizations in the world to build prototype boats for salt spraying is not a climate or governmental body but the Discovery television channel, to make an entertaining documentary.[15]

Why are governments reluctant to spend money on geoengineering? Partly it would seem, because of the complex international political issues that would arise should any nation pursue the technology. But it is also partly because many climate change campaigners trenchantly oppose geoengineering. The activists would rather we cut carbon dioxide emissions at any cost than invest in a solution that could allow factories to continue belching carbon dioxide into the air. The campaigners are less concerned with reducing the rise in temperature than they are with reducing the use of fossil fuels. This seems unreasonable.

Of course, just avoiding the temperature rise wouldn't solve every climate-change-related problem. But remember, no realistic climate policy promises to solve all (or even most) climate change problems.

If geoengineering could avoid a significant fraction of expected climate change damages with few or no side effects, it would be a phenomenally useful intervention for humanity. Indeed, given that today we're contemplating carbon-cut policies that will cost hundreds of trillions of dollars yet do little to help, paying instead $0.009 trillion ($9 billion), as Copenhagen Consensus research suggests, to fix a significant part of climate change would leave an enormous amount of money available to do good in other ways.

The researchers working for Copenhagen Consensus who studied the merits of geoengineering recommend starting research spending with tens of millions of dollars now, and increasing this to the low billions in decades. This would be a sensible level of investment to make sure that geoengineering in fact can work, will generate as few downsides as possible, and all of us are well informed.[16]

We should *not* commence geoengineering now, since the technology is not ready and we don't yet know enough about it. But we simply can't afford *not* to research it. It might just prove to be the earth's best backup plan.

PROSPERITY:
THE OTHER CLIMATE POLICY
WE NEED

WE HAVE LOOKED at how to set a carbon tax, at what we can achieve with innovation and adaptation, and at the backstop policy of geoengineering. But there is one other approach that doesn't get enough attention, and in fact isn't even commonly thought of as a climate change policy. It's making countries richer.

The case for seeing prosperity as climate policy is made clear when we look at two low-lying nations built on river deltas, Bangladesh and the Netherlands, and examine the differences in their exposure to climate change. Both countries are flood prone: 60 percent of Bangladesh is vulnerable to flooding, and 67 percent of the Netherlands is similarly vulnerable. Both face considerable challenges from global warming. Yet their vulnerability to climate change, and response to it, is very different. This is obvious when looking at flooding from rising sea levels.[1]

The Netherlands suffered a devastating flood in 1953, the Watersnoodramp. More than eighteen hundred people died when water breached the dikes in the provinces of Zeeland, South Holland, and North Brabant. The disaster prompted the Dutch government to invest massively in flood prevention over the next half century, building an extensive system of dams and storm-surge barriers. This program, the Delta Works, ended up costing about $11 billion in total. It was such an enormous project that it was sometimes referred to as the eighth wonder of the world. Since 1953, there have been just three floods and one fatality due to flooding in the Netherlands.[2]

In contrast, Bangladesh still sees large-scale flooding. In 2019, flooding forced more than two hundred thousand Bangladeshis to flee their homes, and prompted fears of food insecurity for four million. In the first two decades of the twenty-first century, more than three thousand people have died from flooding. Every year, flooding causes devastation and claims lives.[3]

The obvious truth is that rich countries can spend more money to protect against climate change than poor countries can. Lifting countries out of poverty is an essential but underdiscussed approach to mitigating the damage of climate change.

Let's take another look at three of the pathways developed by scientists working for the UN (from chapter 9). In each scenario, Bangladesh is set to become richer. If we look at the middle-of-the-road pathway, it will by the end of the century be as rich as Holland is today. If we look at the best scenarios—sustainable (green) and fossil fuel—which will see large-scale investment in education, health, and technology, then by the end of the century Bangladesh will be much richer still.

In the sustainable scenario, Bangladesh will in the early 2080s become richer than today's Holland. In the fossil fuel scenario, it will become even richer, even faster, surpassing today's Holland in the late 2060s, less than fifty years from today.

As Bangladesh gets richer, it will be able to spend more on adaptation. Flood defenses for roads and railways, river embankments protecting productive agricultural lands, and drainage systems and erosion control measures for major towns are estimated to cost almost $3 billion initially and $54 million in annual costs to 2050. While the initial cost is about one percent of Bangladesh's GDP today, it will be one-tenth or less in the 2050s. Clearly, widespread adaptation will follow as Bangladesh gets richer, and more adaptation will follow with more development. Before the end of the century, it seems likely that if Bangladesh becomes richer than Holland is today, it will have flood and sea defenses at least as good as those now found in the Netherlands.[4]

Today, Bangladesh spends almost $3 billion annually to subsidize fossil fuels like gas and electricity production as a way to provide citizens with

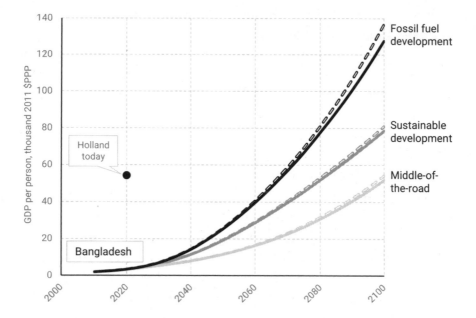

FIGURE 15.1 Three scenarios for GDP per person in Bangladesh over the century. Dashed lines show the actual amount; full lines show the amount when adjusted for climate damages. Holland in 2020 is shown for reference.[5]

energy access. As incomes rise, it is likely Bangladesh can not only cut its subsidies, but start taxing carbon emissions and begin spending on green energy R&D like richer countries already do. If you're poor, you burn cheap, dirty fuel. If you're richer, you can afford to subsidize wind turbines.[6]

Picking the high-growth pathways to ensure that Bangladesh gets out of poverty and into a better life with more capacity for adaptation, innovation, and green taxes *is in itself* a climate policy. Yes, development will also lead to more carbon emissions, higher temperatures, and greater climate damages, as is indicated as the difference between the dashed and solid lines in figure 15.1. But the positive effects will vastly outweigh the negative ones, in terms of higher climate resilience, more sustainable long-term climate policies, and in terms of development itself.

THE REASON WHY we pursue any climate change policy is to make the world better than it would otherwise be. Our goal is to make sure that both people and the world's environment are better off than they would have been had we done nothing. This is why we set a carbon tax to ensure that emissions are lower and hence temperatures do not rise as high. We invest in green R&D to find low- or zero-carbon energy technologies to outcompete fossil fuels, lower emissions, and reduce temperature rises. We invest in adaptation and geoengineering to reduce and counterweigh the negative impacts of global warming.

But it is an unavoidable fact that all these policies also cost us resources that could have been spent making people's lives healthier, longer, and more prosperous. If we focused some of these resources on effective development and human capital investments, then people would be more able to afford expensive green energy sources and more capable of investing in adaptation. As a society, we would have more money to respond to climate change.

This is especially true for the world's poorest. As we saw in chapter 10, climate change affects the world's poor much more than it affects the rich. Indeed, often the climate travails of the world's poorest are used as an explicit reason for cutting carbon dioxide emissions and diverting development aid to climate projects.

In countries most affected by spikes in temperature, it is very clear that getting poor people out of poverty means they will not be nearly as affected by heat waves. Their societies will be better able to afford air-conditioned hospitals and community centers that can help especially vulnerable people cope with the heat. It means their societies can afford to make urban environments cooler, with more parks and water features, cool roofs, and cool pavements. It also means individuals can increasingly afford to keep cool when it's hot and keep warm when it's cold, because of the increased affordability of cooling and heating.

Ensuring more widespread prosperity also means fewer people will be dependent on small-scale farming, in which a single weather catastrophe can spell disaster for a whole family. It can help societies change from being mostly dependent on weather-affected agriculture to a much more secure life within weather-proof manufacturing and services. A more prosperous society will also have better health services, which can help ameliorate negative impacts from global warming. They will have better social protections, making sure fewer are affected by nutritional deficiencies, even if climate change harms local agriculture.

History shows, too, that making a society richer means it can afford to stop deforestation. The Netherlands is planting forests while Bangladesh is still cutting them down. As people move out of poverty and don't have to focus so much on simply surviving, they become more interested in safeguarding nature, cutting air and water pollution, and starting to reforest. While the area of global wetlands and its biodiversity is expected to decline as sea levels rise without extra resources spent on wetland adaptation, richer societies will be more likely to set aside resources to create new space for wetlands. In total, richer societies will likely see up to 60 percent *more* wetlands.[7]

And of course, a society that is better off enjoys many other improvements that go far beyond being able to deal with climate change. Young people will have better education and more job opportunities, and more money can be invested in health care and protecting the vulnerable. The Netherlands has its problems, but extreme poverty is not among them.

This does *not* mean we should do nothing to address climate change. There are plenty of climate policies that are so cost effective, we should definitely implement them. But it does mean we need to more explicitly weigh the costs and benefits of climate policies against all other policies and ask, where can we help most? We will discuss answers to this broader question in the next chapter, but not surprisingly, it turns out that one of the best answers is to increase the level of well-being in the Bangladeshes of the world, as fast as we can.

THE IDEA OF prosperity as a climate change policy is rarely considered. We are much more focused on specific actions like putting up a solar panel or abstaining from meat. But the idea has been around for decades. Back in 1992 when climate negotiations were just beginning, Nobel Prize–winning economist Thomas Schelling (with whom I collaborated for decades) first posed the question, are poor people really best helped through cutting CO_2 and adaptation, or could we achieve more if we focused on making them more prosperous? The so-called Schelling conjecture suggests that getting richer is likely to be the better way to help people, even those faced with climate problems.[8]

One of the clearest examples of an answer to the Schelling conjecture is found in a 2018 study that followed sixteen hundred rural households in Tanzania over six years, analyzing their vulnerability to weather shocks. The study found that hotter years affected the very poor differently than it did others. The very poor households had less food and opportunity for consumption in hotter years. But those who were one rung or two higher on the ladder, who were still poor but slightly less so, turned out not to be affected by weather shocks. Their food and overall consumption possibly even *increased* slightly in hotter years. Why? These "slightly less" poor have multiple (small) income streams, not just from agriculture, but often also from business and retail. They can vary their incomes better when temperatures spike. They can also better afford to invest in things like irrigation, or maybe take a chance on higher-yielding but riskier seed varieties. In comparison, the absolutely poorest will understandably shy away from such risky investments, and will be confined to back-breaking outdoor agricultural work that becomes harder in temperature spikes.[9]

The lesson is simple: even if you just want to help the poorest people in Tanzania reduce their vulnerability to climate change and nothing else, the best way to help them is to lift them out of extreme poverty. Get the poorest people to become, at the very least, slightly less poor.

Indeed, a 2012 study investigating the impact of climate and prosperity policies for all regions of the world shows that even when the goal is just to help with climate impacts, for the poorest regions the best policies focus on development, not climate. Even if our singular goal is to reduce the impacts of climate change, one of the best ways to assist the world's poorest is to help them escape poverty. Prosperity can be a highly effective climate policy.[10]

TACKLING CLIMATE CHANGE AND ALL THE WORLD'S OTHER CHALLENGES

CONCLUSION:
HOW TO MAKE THE WORLD
A BETTER PLACE

THE GOAL OF climate policy is to make the world a better place. Right now, we stand at a crossroads. We can charge on in the same direction, but thirty years of failed climate policy tell us this approach will not make the world much better, and it will cost us a lot. Or we can choose a different pathway that could help people and the planet much more.

Let's look at the big picture for a moment. Instead of thinking just about the single issue of climate change, let's look at how well humanity is doing right now on *all* of the biggest challenges confronting us.

Are things getting better or worse? One way of answering this question is to look at these challenges, expressed as costs to humanity.

Figure 16.1 is the result of work I undertook with ten world-class teams of economists. We set out to establish what different global problems have cost—and will cost—from 1900 to 2050, as a percentage of global GDP. This figure reveals two essential points.[1]

First, things are generally getting better. From air pollution to gender inequality to malnutrition, we have managed to reduce the impact of the biggest challenges humanity has faced, and we are on target to reduce them even further. If we think about what the world looked like in 1900 and what it looks like today, this lesson comes as no surprise.

Second, the figure shows us that climate change is a moderate problem in a sea of problems, big and small.[2]

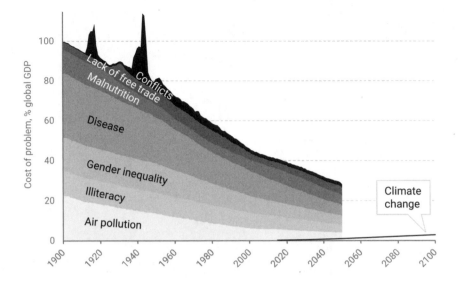

FIGURE 16.1 How much richer would the world have been had we solved different issues, 1900–2050. Cost of climate change with optimal policy added for reference.[3]

These two points need to guide us. We have reduced the scale of other challenges, and we need to continue that work for the future. We also need to solve global warming. But we need to do so with the knowledge that climate change is not the only problem facing the world, nor is it the largest.

PEOPLE ARE TERRIFIED of climate change, above all else. Half the world's surveyed population believes human extinction is likely, and that faced with possible annihilation, any expenditure is justified.[4]

But remember, we have been scared witless before. In the 1960s and 1970s, the good and necessary concern for the environment took a dark turn. Thought leaders were supremely confident that the planet faced doom. They foretold a future of overpopulation, famine, and pollution, with social and ecological collapse.

No one person did more to preach this pessimistic message than Stanford ecologist Paul Ehrlich. He was a favorite on Johnny Carson's *Tonight Show*, prophesizing to the American public that the planet was on the brink of destruction, not least because ever more people were gobbling up resources and generating pollution. On CBS national news in 1970, he explained that the world would end before 1985: "Sometime in the next fifteen years, the end will come. And by 'the end' I mean an utter breakdown of the capacity of the planet to support humanity." Other influential academics agreed that the end was near, differing only in how much time was left: as little as five years or as many as thirty. The media jumped on these stories. *Life* magazine, for instance, reported in 1970 that "horrors lie in wait," and that within ten years "urban dwellers will have to wear gas masks to survive air pollution."[5]

But the warnings were completely wrong. Ehrlich predicted that one to two billion people would die from starvation in the 1970s. He was off by much more than 99 percent. He claimed that by 1980, the life expectancy of the average American could be forty-two years. In 1980, it reached seventy-four years. Los Angelinos didn't need gas masks for air pollution but instead breathed easier in the 1980s because of technological breakthroughs like the catalytic converter and stricter environmental regulations. The world did not end.[6]

The apocalyptic rhetoric did significant damage, however. If the world was becoming overpopulated, the conclusion drawn by researchers and politicians was that it was necessary to make people have fewer children, especially the poor. Population control came to be seen as "the only possible salvation of the underdeveloped world," as Ehrlich put it. Some researchers started considering adding chemicals to water supplies or staple foods to make the world's poor temporarily sterile. It gave urgency and legitimacy to end horrific abuses committed by governments, including forced sterilization campaigns and coerced abortions. In just one year, in 1976, the Indian government force-sterilized 6.2 million men.[7]

And the idea that the world was on the brink of irrevocable collapse from starvation even made it possible for some leading environmentalists to consider cutting off food aid to entire countries and regions. Believing that continued food aid would allow the births of more children, who would inevitably die in later hunger catastrophes, Ehrlich advocated that humanity's most hopeful scenario was to cut off food aid to Vietnam, Thailand, Egypt, and India—and sit back and watch while famine and food riots killed half a billion people. It is horrific to consider the human suffering that would have resulted, had we not instead followed the advice of Norman Borlaug and the green revolution and pursued a pathway of innovation and improving agricultural productivity. Ehrlich's approach is possibly the most heinous example in modern history of bad policy advice emerging from unfounded, apocalyptic worries.[8]

PANIC DOESN'T JUST lead us toward bad or ineffective policy solutions—it can also lead us to focus on the wrong problems. This is a challenge we've seen throughout the history of the environmental movement.

Much of the early concern of environmental organizations, campaigners, and politicians was with regulation to protect us from potentially dangerous toxins. In large part, this focus was due to the outsized impact of Rachel Carson's 1962 bestseller, *Silent Spring*, which made the case that chemicals like DDT and other toxins were killing birds and children. Carson was correct in pointing out that the US government was not doing nearly enough to safeguard nature and humans from dangerous chemicals. But even at the time, her concerns were exaggerated; she feared DDT and other toxins were killing ever more children and adults from leukemia, although US death rates from the disease were declining from 1960. Even the United Nations claimed that DDT and other toxins were causing more than *half* of all cancers, although the correct number was probably lower than 2 percent.[9]

The resulting nationwide fear led to a large federal focus on reducing toxin risks. The total cost of regulating toxins ran to more than $200 billion per year by the mid-1990s. But these policies had a very poor return on investment. A research team from Harvard University evaluated the regulations and found that the majority of them were very inefficient. Benzene emission controls at rubber tire manufacturing plants, for instance, cost $100 million per year, but would save a life only once every four thousand years.[10]

The Harvard research team compared toxin controls with other regulations saving lives through injury control programs (such as flammability standards for upholstered furniture) and medical programs (such as cervical cancer screening).

Across 185 life-saving regulations with sufficient data, the total annual cost to America was $21 billion and the regulations saved sixty thousand lives each year. However, if the focus had been on the most effective regulations—much less on toxin controls, more on injury control, and much more on medical programs—the same total cost of $21 billion could have saved one hundred twenty thousand people each year. Put bluntly, the outsized toxin fear of the 1960s and 1970s ended up committing at least sixty thousand annual statistical murders; that is, lives lost that could have been saved had public policy been focused on implementing the most effective regulations first.[11]

We need to stop repeating history. Today's climate change movement explicitly calls for the largest mobilization of resources in history, and for putting humanity on a "war footing" to combat rising temperatures. Rather than making fear-based, knee-jerk policy decisions, we need to make sure that we respond effectively and efficiently.

WE HAVE ALREADY discussed many of the factors leading to simplification and scaremongering in the climate debate. Researchers find it simpler to ignore complex aspects like humanity's adaptive ability when they study the effects of climate change. The media moves swiftly and likes to tell a simple, clear, terrifying narrative. Politicians gain votes when they present themselves standing with voters against an external threat. Everyone wins.

This alignment of interests has led us to bad choices in the past. You've probably heard of the "military-industrial complex," a phrase coined by President Eisenhower in 1961 to warn against the undue influence of an immense military establishment and arms industry creeping into "every city, every State house, every office of the Federal government."[12]

The military-industrial complex arose during the Cold War as a response to the perceived threat of Soviet strength. In the 1950s, the conventional wisdom was that the Soviet Union was experiencing incredibly fast economic growth and was about to eclipse the United

States. The threat was used as a powerful argument for much more defense spending.

Statistics on Soviet growth turned out to be dramatically exaggerated. Yet, at the time it was hard to stand up and say, "Maybe we should spend more on health and education instead?" without being labeled a communist sympathizer. And so the military grew and grew, along with the industries that supported it and the universities that provided military-related research. Multiple forces conjoined to create a national panic. Children performed air raid drills in school, people built bunkers in their backyards.[13]

The costs of the panic were immense, not just in actual resources spent, but also in the amount of forgone opportunity to invest elsewhere. What if the money poured into America's massive military had gone instead to cancer research? To education? To prenatal care for poor mothers? To rebuilding infrastructure? What would America, indeed the world, look like now had it not been for this false alarm?

SOMETHING AKIN TO the military-industrial complex is emerging surrounding climate change, though some of the players are different this time around. Climate change needs our attention, but it has become an all-consuming focus, in part because so many people benefit from the panic. Politicians and media executives benefit, of course, but so does big business.

No doubt, many CEOs are genuinely concerned about global warming, but they also stand to make a great deal of money from carbon regulations. This interest is particularly obvious in the energy sector. Go to any climate conference and you'll find wind energy manufacturers sponsoring media segments on the "climate crisis" that increase support for policies that would significantly add to their own earnings—while urging governments to invest heavily in the wind market.

Even companies not heavily engaged in green energy stand to gain. Some businesses blatantly wrap themselves in a bright-green mantle for the sake of branding, of course. But others stand to gain in ways that aren't always clear to consumers. In the first decades of European climate policy, for instance, many energy companies made billions of dollars in extra profits from Europe's cap-and-trade system. The European Union intended for energy companies to buy certificates to counteract all their emissions, and the businesses would then pass the cost on to consumers, thus creating a financial incentive for company and consumer alike to reduce fossil fuel reliance. However, it is much easier to get companies to accept legislation if it makes them money rather than costs them money. So in practice, European governments gave most of the certificates to the companies free of charge, but the companies continued to charge their consumers as if they had paid for them. In just the first eight years of EU emissions trading, this made companies, including many coal-fired power plants, about $80 billion in additional profits. US energy companies had high hopes that they could benefit similarly when the United States was considering cap-and-trade legislation in 2009. Energy companies' lobbying costs *for* climate change action more than tripled to over $350 million for that year.[14]

Enron pioneered the cozy corporate relationship with politicians to support climate change action. It bought up renewable energy companies and carbon-credit-trading outfits, with the expectation that they would gain significantly in value from climate agreements. When the Kyoto Protocol was signed, a memo written within Enron stated: "If implemented, [the Kyoto Protocol] will do more to promote Enron's business than almost any other regulatory business." The moral of the story? Especially when a company is calling for more environmental regulation, we need to look very carefully at what they might stand to gain from it.[15]

Environmental campaigners also play a part in this story, sometimes unwittingly, by helping to stoke fear. Enron boasted of their "excellent credentials with many 'green' interests" including Greenpeace and the Climate Action Network. This relationship boosts companies' bottom line, because campaign organizations help increase public willingness to pay for subsidies and other handouts to fund expensive climate policies, through visions of catastrophe and Armageddon.[16]

I don't believe that there is some kind of grand conspiracy to promote scare stories about environmental crisis. I do believe that companies, the media, and politicians benefit from those stories. This confluence of interests goes a long way to explaining why the conversation surrounding climate change has become so detached from scientific reality.

WHAT IS THE POINT of climate change policy? To make the world a better place for all of us, and for future generations. To me, that suggests that we need to ask ourselves a broader question. If the goal is to make the world a better place, is climate change policy the most important thing to focus on?

Certainly, it is one thing we should focus on. We must rein in temperature increases and help ensure that the most vulnerable can adapt. But today's popular climate change policies of rolling out solar panels and wind turbines have insidious effects: they push up energy costs, hurt the poor, cut emissions ineffectively, and put us on an unsustainable pathway where taxpayers are eventually likely to revolt. Instead, we need to invest in innovation, smart carbon taxes, R&D into geoengineering, and adaptation.

But we also need to recognize that reducing global warming is only one of many things we can do to make the world a better place.

Making the world richer is also important. If we end up on a pathway that creates the highest possible welfare for the planet, we could leave the world $500 trillion better off each and every year by the end of the century. This is a virtuous goal in and of itself, but it will also help address climate change. The richer people are, the more resilient they will be in the face of global warming.

To make the world richer, we must invest in health care, in education, and in technology. But we can't do it all. Current climate policy is so expensive, and will drain so much of potential future gains in GDP, that it will leave less money for policies that will enhance prosperity.

Ultimately, there is a limited amount of money we can spend on making the world a better place. So we need to make choices, and we need to make trade-offs. The good news is that we have a great deal of data indicating what the best investments are.

The United States, together with almost all other countries in the world, has adopted the so-called Global Goals or Sustainable Development Goals. These were set by the United Nations and list 169 targets encompassing everything from reducing gender-based violence and poverty, to increasing nutrition and broadband access, to fixing climate change and poverty.

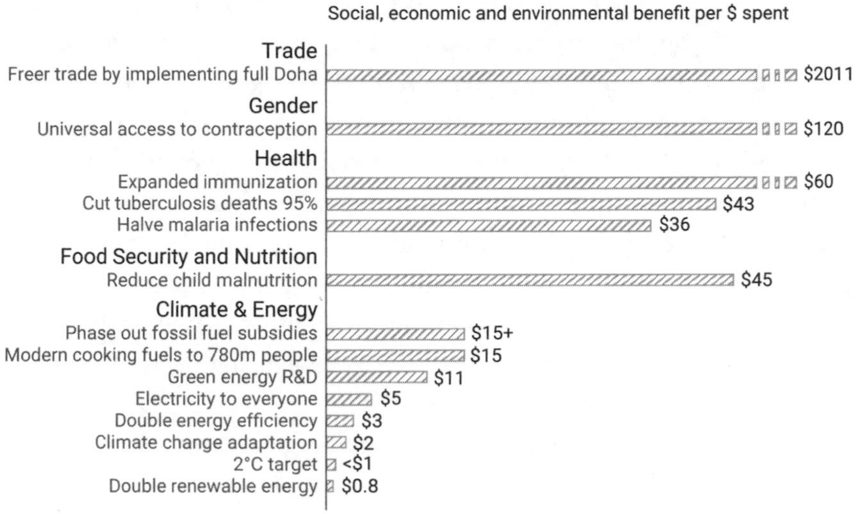

Social, economic and environmental benefit per $ spent

Trade
Freer trade by implementing full Doha — $2011

Gender
Universal access to contraception — $120

Health
Expanded immunization — $60
Cut tuberculosis deaths 95% — $43
Halve malaria infections — $36

Food Security and Nutrition
Reduce child malnutrition — $45

Climate & Energy
Phase out fossil fuel subsidies — $15+
Modern cooking fuels to 780m people — $15
Green energy R&D — $11
Electricity to everyone — $5
Double energy efficiency — $3
Climate change adaptation — $2
2°C target — <$1
Double renewable energy — $0.8

FIGURE 16.2 How much do different policies help the world? Each policy has been analyzed by economists to estimate its total costs and total benefits. The bars on the right-hand side show how much social, economic, and environmental benefits one extra dollar will achieve—longer bars are better.[17]

My think tank, the Copenhagen Consensus, worked with fifty teams of economists and several Nobel laureate economists to analyze these development investments and find which initiatives will achieve the greatest "return on investment" for humanity (see figure 16.2).[18]

As we have throughout this book, we can think about these policies in terms of costs and benefits. Smart climate policies give a good return on investment. But we can achieve much more for humanity with interventions in areas other than climate policy.

Top among these is freer trade. Free trade has recently been criticized by left- and right-wing politicians because it hurts vulnerable communities like manufacturing workers in "rust belts." This misses the bigger picture. Governments could spend billions more supporting the most vulnerable communities that would be hurt by free trade deals, and benefits would still vastly outweigh costs. Globally, freer trade could unleash an amazing $2,000 of social benefits for every dollar spent. Much of these benefits would go to the world's poorest, who would have far more opportunities if they could become part of the global market. Unfortunately, the world has almost given up on the latest round of trade negotiations, called the Doha Development Round, but if we were to successfully conclude such a global treaty on freer trade, economists estimate that it could by 2030 make the average person in the poor world a thousand dollars richer per person per year.[19]

And remember, by making people richer, especially in the world's poorest countries, freer trade would also lead to societies far more resilient to climate shocks, more capable of investing in adaptation, and far less vulnerable to rising temperatures. In that way, free trade can be considered a smart *climate* policy as well as an excellent way to promote human thriving generally.

Other policies, too, would be incredibly beneficial for the world *and* help us address climate change by making people better off, and hence more resilient. Avoiding malnutrition in the first two years of a child's life costs about $100. Because good nutrition helps develop the child's brain, it leads to better educational outcomes and phenomenally higher productivity in adulthood. Indeed, this $100 spent will on average increase each child's lifetime income to the equivalent of a onetime amount of $4,500, in today's dollars. Every dollar spent on fighting early childhood malnutrition results in $45 of social good. This research was cited when governments pledged more than $4 billion for nutrition investments, but much more is still needed.[20]

There is a compelling case, too, to spend more resources to fight the world's leading infectious disease killer, tuberculosis. This disease is hugely overlooked by philanthropists and governments. It mostly kills adults in their prime, leaving children without parents. For about $6 billion annually, we could save nearly 1.6 million people from dying each and every year. Parents could continue to work, and children would not be orphaned. When the Copenhagen Consensus undertook an analysis for philanthropist Bill Gates, he concluded that his investment in tuberculosis, immunization, polio, and malaria was the "best investment I've ever made."[21]

Closing the contraception access gap would also be a transformational achievement. There are 214 million women of reproductive age in poor countries who want to avoid pregnancy, but do not have access to modern contraceptive methods. Solving this challenge will cost just $3.6 billion more each year than what is spent today. That will immediately save lives, avoiding 150,000 mothers dying in childbirth each year, because of fewer pregnancies. Better spacing of children from family planning also allows parents to assign more resources of money, time, and emotional energy to each child. This parental investment will make children more productive as adults. As there will also be fewer adults in any age group, each child will eventually also be able to use more capital resources, and entire societies will become more productive, with greater economic growth. Each dollar spent on contraception and family planning education will generate $120 of social benefits across the most vulnerable societies. This research recently helped convince the British government to pledge £600 million ($761 million) to provide 20 million more women and girls in the developing world with access to family planning. While great, this is only a drop in the ocean of what is needed.[22]

What do all these investments have in common, other than being compelling? The fact that they are underfunded. We could go far further in solving each of these challenges if we could allocate more resources. Remember that approximately one-fourth of aid today is diverted to climate aid projects. Indeed, the amount spent on climate aid today could fund all these interventions in contraception, tuberculosis, malaria, immunization, and nutrition—with money to spare. The money spent on climate aid is money we cannot spend twice. Many people deeply committed to addressing climate change also believe in expanding women's access to contraception, in reducing poverty, and in eliminating disease in the poorest parts of the world. But too often, they forget that there are trade-offs.

WE DEFINITELY NEED to address global warming—it's a real problem and we need to put policies in place both to limit its extent and to enable the world to best manage its impact.

But if we truly want to make the world a better place, we have to be very careful that our preoccupation with climate change doesn't distract us from other crucial problems. We can improve the human condition far more by opening the world to free trade, ending tuberculosis, and ensuring access to nutrition, contraception, health, education, and technology.

Fixating on scary stories about climate change leads us to make poor decisions. As individuals, we feel compelled to transform our lives, in ways both minor (not eating meat) and major (foregoing parenthood). As societies, we are making treaties that promise to squander hundreds of trillions of dollars on incredibly inefficient carbon-cutting policies.

Overspending on bad climate policies doesn't just waste money. It means underspending on *effective* climate policies and underspending on the opportunities we have to improve life for billions of people, now and into the future. That's not just inefficient. It's morally wrong.

Over the past century, the world has become a better place, thanks to human ingenuity and innovation. Our choice now is whether we want to allow fear to drive our choices, or if we want to use our ingenuity and innovation to make sure that we leave future generations the best world possible.

ACKNOWLEDGMENTS

Writing a book is really leaning on the bright ideas of so many people who work to further a great world. Thanking them individually would be impossible, but collectively they have made our world so much richer.

I've spent the last two decades working with some of the best economists across the world to identify smart solutions to the world's big problems. I've learned from them all and thank them deeply.

Within climate science, I want to say thanks to the tens of thousands of scientists who have invested an enormous amount of energy into understanding the physical workings of our world, publishing their findings throughout the academic literature, and to the thousands of scientists who have worked to summarize this information in the UN Climate Panel reports.

But to tackle climate smartly, we not only need to know about the physical reality of climate change. We also need to understand its impact on societies when taking adaptation into account, and we need to understand the different social strategies available, along with their costs and benefits. Here, climate economics is invaluable, and I want to thank the large cohort of academics and analysts who have made us all know more. Many of you are widely cited in this book.

In my own organization I'd like to thank the Copenhagen Consensus Center chief economist, Brad Wong, for many educating conversations across almost all conceivable topics, from government procurement in Haiti and tuberculosis treatment in Ghana, to solar panels and disaster preparedness in Bangladesh. I also want to say thanks to David Cooper for helping me prepare this book—you made almost every sentence better! Thanks to all you guys who helped me create this book, and also make it better: David Lessmann, Nancy Dubosse, Saleema Razvi, Ralph Nordjo, Justin De Los Santos, Cyandra Carvalho, Krisztina Mészáros, Loretta Michaels, and Scott Calahan. Roland Mathiasson deserves his own mention—you helped me every step of the way, making the graphs, the arguments, and the book much better. Thanks!

But without my editors at Basic Books, this book would be much less sharp and clear. Thanks go first and foremost to Lara Heimert, who has consistently pushed me through rewrite after rewrite to make my text the best possible version of the argument. Yes, there were times when it felt like I was banging my head against a wall, but you really made my book much, much better. Thanks, Lara!

So many others from Basic Books have been helping to improve the book. Thanks to Roger Labrie for consistently pointing out how my mumbling can be transformed into clear text, thanks to Christine Marra for racing the book through production, to Gray Cutler for a great copy edit, thanks to Kait Howard and Victoria Gilder for the media outreach, and to the many others who worked on the book's behalf: Jessica Breen, Allison Finkel, Katie Lambright, Abigail Mohr, and Melissa Raymond.

At the end of the day, writing can be a lonely endeavor. But all of you helped me make this book better. Of course, any errors are mine and mine alone. But many of the great points, good arguments, and insights are thanks to all of you.

Prague, February 25, 2020

NOTES

INTRODUCTION

1. **News outlets refer:** Roberts, 2019; **planet's imminent incineration:** Hodgetts, 2019; **Recently, the media:** Holthaus, 2018; Climate Nexus, 2018.

2. **Activists are cordoning:** Extinction Rebellion, 2019; **"slaughter, death, and starvation":** Climate Feedback, 2019b.

3. **In 2017, journalist:** Wallace-Wells, 2017; **Although the article:** Climate Feedback, 2017, and Wallace-Wells, 2019a; **It could finish:** McKibben, 2019b. *Field Notes from:* Kolbert, 2006; *Storms of My:* Hansen, 2011b; *The Great Derangement:* Ghosh, 2017; *This Is the:* Nesbit, 2019.

4. **"across the globe . . .":** Oreskes, 2015; **Be worried:** *Time,* 2006; **The British newspaper:** *Guardian,* 2016; **"sounds rather passive . . .":** Carrington, 2019.

5. **In the United Kingdom:** Dahlgreen, 2016; **In the United States:** M. Smith, 2019.

6. **"I know that . . .":** Astor, 2018; **"How Do You . . .":** Ostrander, 2016.

7. **A 2019** *Washington Post:* Kaplan and Guskin, 2019; **A 2012 academic study:** Strife, 2012.

8. **Recently, a Danish:** Henriksen, 2019; *Parenting in a:* Berrigan, 2019; *On Having Kids:* Braverman, 2019.

9. *The Skeptical Environmentalist:* Lomborg, 2001; **Scientists agree that:** Indeed, the total impact estimates of climate change, the so-called social cost of carbon, has been *declining,* not increasing from 1996 till today, suggesting damages are expected to be lower, not higher (Tol, 2018, 14).

10. **Climate change is:** I'll use climate change and global warming interchangeably in this book, although climate change has a somewhat broader meaning; **"We risk turning . . .":** J. Smith, 2019.

11. *This is what:* J. Watts, 2018.

12. **In the United States:** National Safety Council, 2019; **If politicians asked:** Remarkably, the world's first traffic death occurred at four miles per hour, so better set this limit at three miles per hour (Guinness 2019).

13. **In 1900, the:** Roser, 2019a; **The world is:** Roser and Ortiz-Ospina, 2019b, Ortiz-Ospina and Roser, 2019; Roser, 2019b; **Between 1990 and:** H. Ritchie and Roser, 2019b. **In 1990, pollution:** Institute for Health Metrics and Evaluation, 2019; **Higher agricultural yields:** Ewers, 2006; **And since 1990:** World Health Organization, UNICEF, and WHO/UNICEF Joint Water Supply and Sanitation Monitoring Programme, 2015, 7f.

14. **Over the past:** World Bank, 2019c; **Today, it is:** Roser and Ortiz-Ospina, 2019a.

15. **Researchers working for:** According to the UN's middle-of-the-road scenario, which we will discuss later (IIASA, 2018); **Life expectancy will:** United Nations, 2019b; Lutz, Butz, and KC, 2014, 247.

16. **The best current:** W. Nordhaus, 2018.

17. **For most economic:** IPCC, 2014c, 662.

18. **Evidence shows that:** This is a severe underestimate. Just global subsidies to renewables will reach $176 billion in 2020 (IEA, 2018, 256) and the EU climate policy will cost on average 1–2.2 percent of GDP, or $192–$408 billion per year in 2020 (Bohringer, Rutherford, and Tol, 2009).

19. **With 194 signatories:** Lomborg, 2020.

20. **However, it turns:** Lomborg, 2020.

21. **Across the world:** Kotchen, Turk, and Leiserowitz, 2017; Jenkins, 2014; Duan, Yan-Li, and Yan, 2014; **A 2019 *Washington Post*:** Dennis, Mufson, and Clement, 2019.

22. **In the rich world:** IEA, 2017; **For instance, energy:** *Belfast Telegraph*, 2014; **In fact, financial:** Borenstein and Davis, 2015.

23. **Not surprisingly, a:** Campagnolo and Davide, 2019.

24. **the first tons:** Here and everywhere I use "ton" as a metric ton, or 2,205 pounds.

25. **Yet, the extra:** All of these numbers are explained in later chapters.

26. **Top climate economists:** See, for example, the case made by top economists, including three Nobel laureates, on the basis of a large number of research articles from climate economists in the edited work (Lomborg, 2010).

27. **The models show:** Galiana and Sopinka, 2015.

28. **While everyone in:** From 0.06 percent of GDP around 1980 to 0.024 percent in 2017 (IEA, 2019c).

29. **A UN global:** United Nations, 2019a.

30. United Nations, 2019a.

31. **It is instead:** T. Nordhaus and Trembath, 2019.

CHAPTER 1.
WHY DO WE GET CLIMATE CHANGE SO WRONG?

1. **instead, we read:** Bowden, 2019.

2. **The actual 2019 study:** Spratt and Dunlop, 2019; **The report presents:** Spratt and Dunlop, 2019, 9; **As one climate scientist:** Climate Feedback, 2019a.

3. ***USA Today*:** E. Weise, 2019; **CBS News:** Pascus, 2019; **CNN:** Hollingsworth, 2019.

4. **Take the June:** *Time*, 2019.

5. **But it would:** Kench, Ford, and Owen, 2018.

6. **The main Tarawa:** Biribo and Woodroffe, 2013; **Similarly, the Marshall Islands:** M. R. Ford and Kench, 2015.

7. **latest research summarizing:** Duvat, 2019.

8. **A similarly scary story:** Lu and Flavelle, 2019; **The headlines stem:** Kulp and Strauss, 2019.

9. From Kulp and Strauss, 2019, and coastal.climatecentral.org. Story from the *New York Times* in Lu and Flavelle, 2019. Because of the way the database is set up, you cannot generate a map for 2020, but only for 2030. So, the difference is really between the lowest emission scenario for 2030 and the highest emission scenario for 2050.

10. **"Climate change is . . .":** McKibben, 2019a; **Climate scientist Peter Kalmus:** Kalmus, 2019.

11. **And it is almost:** Lomborg, 2019. The picture to the left is the one shown in Lu and Flavelle, 2019. The picture to the right is based on Kulp and Strauss, 2019, and their online tool at https://coastal.climatecentral.org. The left-hand picture was generated with RCP8.5, 2050, sea level rise only. The right-hand picture should ideally be the difference between land at risk now and in 2050. Since the Climate Central engine doesn't allow 2020, I've used 2030 with the most extreme climate policies. **In South Vietnam's:** Nguyen, Pittock, and Connell, 2019.

12. **That's fine for:** Goodell, 2019.

13. **As we will:** Lincke and Hinkel, 2018.

14. **One of the most:** IPCC, 2018; **CNN told us:** CNN, 2018.

15. **"policy-relevant but . . .":** IPCC, 2010; **"require rapid, far-reaching . . .":** IPCC, 2018; **Simply put, politicians:** One of the authors of the IPCC report also says the twelve-year limit is "misleading," although it is clear he is much more in line with wanting to cut as much carbon dioxide as possible (Allen, 2019).

16. **This dramatic misrepresentation:** Hulme, 2018. In order to fight climate change, influential environmentalist James Lovelock claims that "it may be necessary to put democracy on hold for a while" (Hickman, 2010).

17. **In the United States:** Egan and Mullin, 2017.

18. **Up until the:** Grunwald, 2019.

19. **Today, people who:** Pew Research Center, 2019.

20. **Republican states have:** T. Williams, 2019.

21. **In the wake:** Thompson, 2019; Savage, 2019; **China has tripled:** IEA, 2019d; **According to official:** IEA, 2019g.

22. **In 2019, Britain's:** Furness, 2019; **he had "calculated . . .":** Verkaik, 2009; **In 2006, Al Gore:** CBS News, 2006.

23. **"win—or lose . . .":** Annon, 1991; **"We all know . . .":** S. Johnson, 2012; **More catastrophic:** Paulson, 2013.

24. **Nearly a decade:** UNEP, 1982; **"devastation as complete . . .":** AAP-Reuter, 1982; **During the 1970s:** Peterson, Connolley, and Fleck, 2008.

25. **"Running Out of . . .":** *Newsweek*, 1973; **"into oblivion":** Ehrlich, 1971, xii.

26. **"to make mankind's . . .":** *Time*, 1972.

27. *The Limits to Growth:* Meadows and Club of Rome, 1972.

28. **Spoiler Alert: They:** Lomborg, 2012.

29. **"sea-level rise could . . .":** *Washington Post*, 2019; **"swallowed by the . . .":** Magill, 2019.

30. **The headlines come:** Bamber et al., 2019; **paper published in 2011:** Nicholls et al., 2011.

31. **In his influential:** Wallace-Wells, 2019a.

32. **This was helpfully:** Simon Davies, 2018.

33. **The authors of:** Jevrejeva et al., 2018; **standards of protection:** Jevrejeva et al., 2018, 8.

34. **Likewise, rapidly growing:** Hallegatte et al., 2013.

35. **And the evidence:** Hinkel et al., 2014.

36. **The highly quoted:** Hinkel et al., 2014.

37. The study shows the number of people flooded from 2000 to 2100 with and without adaptation, in a rich world with very high sea level rise (between two to four feet, or 55–123 cm) by 2100. The upper line shows the large flooding impact without adaptation (when no nation increases its dike heights, no matter how rich it becomes or how much the sea level rises). The lower line shows the flooding impact with realistic adaptation, where all nations increase dikes as sea levels rise, and richer nations increase dike heights even more (Hinkel et al., 2014).

38. **But the total:** Notice that this study relies on estimates of the number of people vulnerable that are possibly underestimated by a factor of three, as was the case in the study discussed above in the *New York Times* (Kulp and Strauss 2019). It is hard to estimate how much it changes the impact, since it is likely that many of the extra people will be protected by the same dikes that Hinkel et al., 2014, are already predicting. At best, the cost is the same. At worst, it will increase the extra dike cost of $24 billion (the difference between the $48 billion under adaptation to the $24 billion without adaptation in 2100) by a factor of three. So, the total cost in 2100 under adaptation would at most run to $38 billion in flood costs, plus $24 billion in basic maintenance, and 3 × $24 billion in extra adaptation (or $134 billion per year); or at most, 0.013 percent of a $1,015 trillion GDP, still way lower than the 0.05 percent spent in 2000.

39. **As a result:** Bouwer and Jonkman, 2018.

40. **"Meeting Paris Climate . . .":** Stieb, 2019.

41. **But here's a:** Lo et al., 2019.; **The city will:** City of Atlanta, 2015.

42. **Adaptations like improved:** R. E. Davis et al., 2003; **France introduced reforms:** Bamat, 2015; **As a result:** *Economist*, 2018; **And Spain cut:** Barcelona Institute for Global Health, 2018.

43. **Well, it turns out:** Heutel, Miller, and Molitor, 2017.

44. Ashley et al., 2014.

45. **Since then, the:** Lomborg, 2020.

46. **This is a:** Strader and Ashley, 2015; Ashley et al., 2014.

47. **One 2017 study looked:** Ferguson and Ashley, 2017.

CHAPTER 2. MEASURING THE FUTURE

1. **"We have known":** Arrhenius, 1896.

2. **Carbon dioxide gas:** Actually, carbon dioxide is not the only gas that warms up the world—so do methane and nitrous oxide (laughing gas)—but it is by far the most important, and often these other gasses will simply be subsumed as so-called carbon dioxide equivalents. **Each year's:** IPCC, 2013a, 486; **So, the amount:** IPCC, 2013a, 11.

3. **One of these:** Meinshausen, Raper, and Wigley, 2011; **Of these, we:** Increasingly, arguments are made that future emissions along a low- or no-policy trajectory will increase less than expected (Burgess et al., 2020; Wallace-Wells, 2019b). This is interesting and, if correct, suggests that the entire challenge of climate change is slightly less dire. It, however, does not change the logic outlined in this book. Indeed, if the

no-policy total emissions are lower, it follows that optimal climate policies would be less drastic than is outlined in the rest of the book.

4. Rich countries are here defined as the members of the OECD, the Organisation for Economic Co-operation and Development.

5. Model based on MAGICC, Meinshausen, Raper, and Wigley, 2011.

6. Model based on MAGICC, Meinshausen, Raper, and Wigley, 2011. In the year 2100 business as usual yields 7.449°F, and without OECD carbon dioxide emissions yields 6.683°F. When rounded, these numbers may seem slightly inconsistent. The graph shows the 2100 temperature of 7.4°F and 6.7°F (difference of 0.7°F), but the actual difference is 0.766°F, rounded to 0.8°F.

7. **They say GDP:** Kennedy, 1968; **Measuring welfare in:** Stiglitz, Fitoussi, and Durand, 2018.

8. **Higher GDP per:** Habermeier, 2007; R. Sharma, 2018; Rosling, 2012.

9. **Global growth in:** Page and Pande, 2018; Dollar, Kleineberg, and Kraay, 2016; **Economic growth has:** Goedecke, Stein, and Qaim, 2018; **And when people:** Steckel, Rao, and Jakob, 2017.

10. **One of the:** Bonjour et al., 2013; **Breathing this foul:** WHO, 2006, 8; **When people emerge:** McLean et al., 2019; **Since 1990, the:** Institute for Health Metrics and Evaluation, 2017.

11. **Put simply, when:** Dinda, 2004.

12. **We see vast deforestation:** Ewers, 2006.

13. **Higher GDP not:** EPI, 2018, figure 3-1.

14. **Even people earning:** Stevenson and Wolfers, 2013, 602.

15. Sacks, Stevenson, and Wolfers, 2012; Stevenson and Wolfers, 2013. Life satisfaction is scored on a satisfaction ladder. Typically asked: Here is a ladder representing the "ladder of life." Let's suppose the top of the ladder represents the best possible life for you, a ten, and the bottom, the worst possible life for you, a zero. On which step of the ladder do you feel you personally stand at the present time?

CHAPTER 3. A FULLER STORY ON CLIMATE CHANGE

1. **However, on a:** IUCN, 2015.

2. **They also survived:** Jakobsson et al., 2010.

3. There is great uncertainty about the absolute number of polar bears, so estimates are typically given as an interval, and here shown as the average between the low and high estimate (IUCN 1986, 63; Wiig et al., 1995 29; IUCN/SSC Polar Bear Specialist Group, 2019, 1).

4. **At that time:** IUCN, 1986, 63. Most of these estimates from the Polar Bear Specialist Group are intervals, from 18,505 to 27,106. For ease of presentation, we here use the midpoint of 22,806 (IUCN Polar Bear Specialist Group and Working Meeting 1985, 40ff). All of the data points are averages between low and high estimates, from IUCN 1986, 63; Wiig et al., 1995, 29; and IUCN/SSC Polar Bear Specialist Group, 2019, 1, which references earlier PBSG estimates from 1997, 2001, 2005, and 2009, along with an updated estimate for 2016, which is presented as "now," here plotted as 2019.

5. **The *Guardian*, a:** Shields, 2019.

6. **The real threat to:** The average number of polar bear "removals" between 2010 and 2014 was 895 (IUCN/SSC Polar Bear Specialist Group, 2019).

7. **When we look:** WWF, 2018; **A 2016 study:** Maxwell et al., 2016.

8. **Scientists who undertook:** Gasparrini et al., 2015.

9. **In one recent:** V. Ward, 2015; **in a single:** Office for National Statistics, 2015b; 2015a.

10. **"Dozens dead in . . .":** CNN, 2019; **Indeed, the biggest:** Fu et al., 2018.

11. Vicedo-Cabrera et al., 2018.

12. **The latest US data:** Vicedo-Cabrera et al., 2018.

13. **And since cold:** Gasparrini et al., 2017.

14. **We don't have:** Heutel, Miller, and Molitor, 2017.

15. **A 2015 analysis:** Diaz et al., 2015.

16. **But global greening:** Mao et al., 2016; **The biggest satellite:** Zhu et al., 2016.

17. **China has seen:** C. Chen et al,. 2019.

18. Using the high RCP8.5 emission scenario: V. K. Arora and Scinocca, 2016; V. K. Arora and Boer, 2014. Vegetation in 1500 from Hurtt et al., 2011.

19. **Researchers find that:** NASA, 2016.

20. **As we emit more:** It also depends on whether the natural environment runs out of other important fertilizers like nitrogen and phosphor; **If we measure:** V. K. Arora and Scinocca, 2016.

21. **But something amazing:** V. K. Arora and Boer, 2014; **By one estimate:** Hurtt et al., 2011.

22. **They argue that:** For instance, Bernie Sanders has made this point (Qiu, 2015).

23. **"an exaggerated focus . . .":** De Châtel, 2014; **"There is very . . .":** Selby, 2019; **Senator Bernie Sanders:** Qiu, 2015.

24. **But as global:** Schlosser et al., 2014.

25. **"the literature has . . .":** Koubi, 2019; **In fact, studies:** D. D. Zhang et al., 2007; **The reason why:** We should be cautious; rising temperatures don't mean that we are off the hook. Much of the data comes from temperate areas like Europe and China, where cold is the bigger challenge. See H. F. Lee, 2018.

26. **More importantly, we:** Mach et al., 2019.

CHAPTER 4.
EXTREME WEATHER OR EXTREME EXAGGERATION?

1. **"A raging, howling . . .":** Achenbach, 2018.

2. **Part of the problem:** Gramlich, 2019; Gallup, 2019.

3. **Amazingly, that's not:** Caspani, 2019.

4. **Indeed, many arguments:** The full claim is, as often, "Ever more frequent and intense droughts, storms and floods." We will tackle the others shortly (UNFCCC 2019).

5. **"There is low . . .":** Stocker and Intergovernmental Panel on Climate Change, 2013.

6. **"drought has decreased . . .":** U.S. Global Change Research Program et al., 2017, ch. 8.1.2.

7. **The US National:** USGCRP, 2017, 236: "There has not yet been a formal identification of a human influence on past changes in United States meteorological drought"; **One 2014 study even:** Hao et al., 2014; N. Watts et al., 2018; **The evidence also:** Donat et al., 2013, 2112.

8. **For the US:** NOAA, 2019.

9. **The UN's climate:** IPCC, 2013b, 1032; **indeed, at a:** Here we're using the fact that the so-called RCP8.5 is extremely unrealistic. See, for example, Wang et al., 2017, and J. Ritchie and Dowlatabadi, 2017; **"is increasingly possible . . .":** U.S. Global Change Research Program et al., 2017, ch. 8.1.3.

10. **In fact, in:** He et al., 2017.

11. **Leonardo DiCaprio's 2016:** Dickinson, 2019; Flavelle, 2019.

12. **"a lack of evidence . . .":** IPCC, 2013a, 112, 214; **The US Global:** USGCRP, 2017, 240. The IPCC 1.5°C report finds that "streamflow trends since 1950 are not statistically significant in most of the world's largest rivers" and that more streamflows are decreasing than increasing (IPCC 2018, 201).

13. **"have not established . . .":** USGCRP, 2017, 231.

14. **In the future:** IPCC, 2018, 203; **"trends in floods . . .":** IPCC, 2013a, 214.

15. **US government scientists:** USGCRP, 2017, 231; **"contribute to increases . . .":** USGCRP, 2017, 242; 2018, 146; **"flood magnitudes . . .":** A. Sharma, Wasko and Lettenmaier, 2018.

16. **It is true:** Lomborg, 2020; **But the number:** Census 2011; 2018a; **Since 1970 alone:** Klotzbach et al., 2018, 1371; BEA, 2019; Census 2018b.

17. Lomborg, 2020.

18. **Every summer in:** "Many consider wildfire as an accelerating problem, with widely held perceptions both in the media and scientific papers of increasing fire occurrence, severity, and resulting losses. However, important exceptions aside, the quantitative evidence available does not support these perceived overall trends" (Doerr and Santín 2016).

19. **The examination of:** Marlon et al., 2008; **To a large:** NAS, 2017, 13.

20. **There is plenty:** Vivek K. Arora and Melton, 2018; F. Li, Lawrence, and Bond-Lamberty 2018; J. Yang et al., 2014; Andela et al., 2017; **And the primary:** Knorr et al., 2014; **In total, the:** J. Yang et al., 2014. 1.4 million km², from almost 5 million km² in the 1900s to just above 3.5 million km² in the 2000s.

21. **Overall, that has:** D. S. Ward et al., 2018, 135.

22. Annual data from Census 1975, L48–55, and NIFC, 2019; decadal data from Mouillot and Field, 2005, 404–5. R. V. Reynolds and Pierson, 1941, table 4, indicates that fire annually consumed even more of the US forests in the nineteenth century; see also Marlon et al., 2012.

23. **"incidence of large . . .":** USGCRP, 2018, 231.

24. **In contrast, climate-related:** Syphard et al., 2017.

25. **Indeed, when we:** Strader, 2018, 557; **Moreover, this growth:** Mann et al., 2014, 447.

26. **When the damage:** Crompton et al., 2010; McAneney et al., 2019.

27. **Compared to the:** Using RCP8.5 and changes in managed lands (Kloster and Lasslop, 2017, 64).

28. **For high-risk California:** Bryant and Westerling, 2014, figure 2.

29. **Hurricanes, scientifically known:** According to global reinsurer Munich Re (Weinkle et al., 2018).

30. **"no significant . . .":** "Current data sets indicate no significant observed trends in global tropical cyclone frequency over the past century" (IPCC 2013a, 216); **They do find:** IPCC, 2013a, 50, 7, 113; **They specifically say:** IPCC, 2013a, 871.

31. **This finding is:** USGCRP, 2017, 259, 258; **Climate scientists at:** They tellingly conclude: "The historical Atlantic hurricane frequency record does not provide compelling evidence for a substantial greenhouse warming-induced long-term increase." GFDL/NASA 2019.

32. **This is true:** Klotzbach et al., 2018.

33. Inflation adjusted with CPI, 2019. Downton, Miller, and Pielke, 2005; Klotzbach et al., 2018; R. A. Pielke and Landsea, 1998, 199; Weinkle et al., 2018, with 2018–2019 from personal communication with Pielke. Dashed lines are linear best fit.

34. **The coastal population:** Data 1900–2010 from Census 1992, 2010, 2012; and 2020 data for the United States from the 2017 population prediction, Census 2017.

35. **From Texas to:** Freeman and Ashley, 2017.

36. **And this isn't:** McAneney et al., 2019; W. Chen et al., 2018.

37. **But as the:** Gettelman et al., 2018.

38. **Hurricane Dorian, which:** D. Smith, 2019.

39. **Currently, according to:** Mendelsohn et al., 2012; updated with essentially same finding in Bakkensen and Mendelsohn, 2016.

40. **The world's best:** EM-DAT, 2020.

41. Data comes from EM-DAT, 2020, using floods, droughts, storms, wildfire, and extreme temperatures for climate-related deaths: average per decade 1920–29, 1930–39, up to 2010–19.

42. **"More billion-dollar . . .":** Brady and Mooney, 2019.

43. **And yes, the:** NCEI, 2019.

44. Costs from 1990 to 2017 from Munich Re in R. Pielke, 2019; 2018 costs from Munich Re, 2019; 2019 costs from Munich Re, 2020, with personal e-mail from Petra Löw, Munich Re, on distribution of geophysical costs. Global GDP from World Bank, 2019e, using World Bank Global Economic Prospects GDP from January 2020 to estimate global GDP for 2018 and 2019. Linear best estimate; decline is not statistically significant.

45. **If we adjust:** Zagorsky, 2017.

46. **It finds that:** Formetta and Feyen, 2019.

47. **For the United States:** AonBenfield, 2019, 40; $39.6B out of a GDP of $20491bn; **Floods have cost:** Arcadia Power, 2014; **Both nationally for:** Formetta and Feyen, 2019.

CHAPTER 5. WHAT IS GLOBAL WARMING GOING TO COST US?

1. **He wrote one:** W. Nordhaus, 1991. There is an even earlier working paper from 1975, W. Nordhaus, 1975, republished after his Nobel Prize: W. Nordhaus, 2019b.

2. Nordhaus and Moffat, 2017. This is an update of the UN's overview (IPCC 2014a, 690, SM10-4). Size of circles shows the weight of the individual studies (larger circles for latest estimates, using independent and appropriate methods; smaller circles for earlier estimates, secondhand studies, or less appropriate methods). The black line is Nordhaus's best estimate, based on median quadratic weighted regression.

3. **Right now, the:** About 1°C, or 1.8°F, since the preindustrial period (IPCC 2018, 51).

4. **Indeed, many impacts:** Although even lower impacts can still mean some cost, because the impacts without global warming would have diminished even faster.

5. **Globally, the value:** Federico, 2005, 233; FAO, 2019.

6. **"Climate Change Threatens . . .":** EEA, 2019; **"Climate Change Could Lead to Food . . .":** B. Johnson, 2019; **"Climate Change Is Coming . . .":** Smyth, 2019; **"Climate Change Could Lead to Major . . .":** Gustin, 2018; **"Climate Change Is Likely . . .":** Little, 2019.

7. **By 2080, in:** Conforti, 2011, 114; **The FAO expects:** Conforti, 2011, 114. Cereal production in 2020 is 2,668 Mt (p. 100); in the 2080s about 3,837 Mt; reduction of 2.1–2.2 percent is 76 Mt less, or 3,761 Mt in the 2080s, or 41 percent higher.

8. **But this is:** Ren et al., 2018.

9. **At the same time:** King et al., 2018; **One study shows:** Challinor et al., 2014.

10. **The biggest study:** Costinot, Donaldson, and Smith, 2016.

11. **The full study:** Costinot, Donaldson, and Smith, 2016. Another study finds total costs of 0.29 percent (Calzadilla et al., 2013). The cost of 0.26 percent is also the conclusion of the recent review of the climate cost on agriculture (Carter et al. 2018); **And actually, this:** This is an average cost for the globe. Poor, tropical countries will do worse and industrialized, cold countries will do better. We'll discuss this general issue in chapter 10.

12. **In the United States:** Weiss, 1992; **Today, it employs:** US Department of Agriculture Economic Research Service, 2019.

13. **In 1991, more:** World Bank, 2019a; World Bank, 2019b.

14. W. Nordhaus and Sztorc, 2013, 11.

15. **While this addition:** W. Nordhaus and Sztorc, 2013, 11.

16. **However, the UN:** W. Nordhaus, 2019a; IPCC, 2013a, 1170; **That is why:** W. Nordhaus, 2019a.

17. **The basic problem:** Seawater is slightly basic, so "acidification" actually means water becomes more neutral, rather than actually becoming acidic.

18. **Economists have tried:** Colt and Knapp, 2016.

19. **This report estimates:** IPCC, 2018, 256.

20. **Yet, when you:** One good example is a recent comment in the *New York Times* (Oreskes and Stern 2019) based on a pamphlet that quoted the 187 million people being displaced from sea level rise as a prime example. As we saw in chapter 1, this number is about six hundred times higher than the actual, likely displacement.

CHAPTER 6. *YOU* CAN'T FIX CLIMATE CHANGE

1. **Carbon intensity is:** Estimating the global carbon intensity, 1966–2017, from IEA, 2019d, using total primary energy supply for all energy.

2. **Since 1992, humanity:** Using data from H. Ritchie and Roser, 2019a; Global Carbon Project, 2019a.

3. **In a surprisingly:** UNEP, 2019.

4. **"must play their part":** UNFCCC, 2018a.

5. **Asked what personal:** BA, 2006; **But if he:** 1.5 w/hr, 12hrs a day = 12 * 0.0015 * 365 = 6.2kWh at 0.527 kg/kWh (Carbon Independent, 2019) or 3.16kg CO_2e, or 0.05% = 3.16/7830kg (World Resources Institute, 2019b); **Charging makes up:** Moss and Kincer, 2018.

6. **It is just:** Narassimhan et al., 2018.

7. **"Don't be distracted . . .":** MacKay, 2010, 114.

8. **In one 2018 study:** Bjelle, Steen-Olsen, and Wood, 2018, using the marginal estimate.

9. **Generally, when researchers:** All examples are based on Bjelle, Steen-Olsen, and Wood, 2018.

10. **If you're making:** Fishbach and Dhar, 2005; **One study of:** Dütschke et al., 2018.

11. **What was most:** Rowlatt, 2007. His entire family carbon dioxide savings was 7.3 tons, whereas the whole family going to Buenos Aires on economy would emit about 18.5 tons of carbon dioxide.

12. **"How about restaurants . . .":** Fagerlund, 2018.

13. **That's especially true:** Humane Research Council, 2014.

14. **Most are vegetarians:** Leahy, Lyons, and Tol, 2010.

15. **Many credulous news:** Martinko, 2014; **This level of reduction:** Hallström, Carlsson-Kanyama, and Börjesson, 2015.

16. **And food-related:** Sandström et al., 2018,

17. **A thorough, systematic:** Hallström, Carlsson-Kanyama, and Börjesson, 2015, table 1; **For the average:** Average carbon dioxide emissions per capita for Annex I is 12.44 tons (World Resources Institute, 2019a), so 4.3%=0.54/12.44.

18. **But there's more:** The literature used to advance the case for cutting out meat clearly says: "When evaluating the environmental consequences of vegetarianism the rebound effect of the savings should be taken into account" (Lusk and Norwood, 2016); **Vegetarian diets are:** Lusk and Norwood, 2016; Grabs, 2015; Berners-Lee et al., 2012; **Spending that extra:** Grabs, 2015.

19. **You could achieve:** Saving 540 kg carbon dioxide, and losing half on rebound, at $6 per ton as of November 2019 (ICAP, 2019).

20. **Artificial meat generates:** Oxford University, 2011.

21. **The International Energy:** IEA, 2019b, 21.

22. **So, switching from:** IEA, 2019b, 21.

23. This is estimated by the European Union (EU, 2019, table 69, 136). The climate cost of 2.1¢ is exaggerated. The EU uses a $110 carbon dioxide cost, which is much higher than its own trading system at $33. Using the cost of actual emissions in 2020, estimated as the average cost across all five SSPs and nine major damage functions, at $20 (P. Yang et al., 2018) gives a more realistic climate damage of 0.4¢. Not shown are the damage costs from gasoline production and transport from well to tank at 0.7¢, and the cost of habitat destruction at 0.9¢.

24. **If you have:** Tessum, Hill, and Marshall, 2014; L. W. Davis and Sallee, 2019; see also https://www.facebook.com/bjornlomborg/posts/10153019443493968:0. Even the

electric car itself emits about as much of the most dangerous air-pollution particulate matter (PM) as gasoline cars (Timmers and Achten, 2016). Electric cars emit no PM from combustion, but because they are typically 24 percent heavier they emit more PM from tires.

25. **It is estimated:** Ji et al., 2012. The extra pollution for the electric cars comes from coal-fired power plants outside the city.

26. **The average subsidy:** Six hundred fifty-five thousand electric cars have cost $2 billion in infrastructure and $5 billion in subsidies, which is a little over $10,000 per car (IEA 2015).

27. **The International Energy:** IEA, 2019b, 120; **Even if we:** IEA, 2019b, 143, estimates a global reduction by 2030 of 220 Mt carbon dioxide, or about 0.4 percent of 2030 global emissions.

28. **Because of the:** Baron, 2019.

29. **But the vast:** Gurdus, 2017; **In India, where:** *Economist,* 2009.

30. **Even if every:** Terrenoire et al., 2019, finds the total impact of the rapidly increasing number of flights across the century to result in a temperature increase by 2100 of 0.1°C. However, this is missing the extra impacts from contrails and aerosols. This is estimated in Olivié et al., 2012, with low and high estimates of contrails and aerosols resulting in a total temperature increase by 2100 of between 0.11°C and 0.24°C. The paper suggests that the most likely real outcome will be somewhere in between, here assessed with the midpoint of 0.174°C. However, most of the temperature increase comes from the many more people who will start flying much more throughout the century, resulting in a more than fivefold increase of total emissions from 2020 to 2100. Cumulative carbon dioxide emissions from 2020 to 2100 is 247 Gt carbon dioxide, using Olivié et al., 2012, table 3. If we assume only the emissions from the current level of passengers, and expecting the same efficiency gain throughout the rest of the century as is expected for all air travel, the total emissions from the current level of passengers flown every year for the rest of the century is about 40 Gt carbon dioxide. The share of the 0.174°C temperature increase due to existing passengers is thus 0.029°C, or 0.05°F.

31. **The rebound effect:** Bjelle, Steen-Olsen, and Wood, 2018; **Sorry, there is:** *Guardian,* 2006.

32. **Over the next:** IATA, 2018.

33. **The overall carbon:** EASA, EEA, and Eurocontrol, 2019.

34. **Billions are being:** See, for example, https://www.cleansky.eu; **And the International Air:** Sullivan, 2018.

35. **"Want to Fight . . .":** Carrington, 2017; **"nearly 40 times . . .":** Galbraith, 2009.

36. **In the 1970s:** Healey, 2016.

37. **The problem with:** See, for example, Murtaugh and Schlax, 2009; **The most cited:** Murtaugh and Schlax, 2009.

38. **Moreover, official expectations:** EIA, 2017b.

39. **And yet people:** And if you still want the cost in money, the amount of raising a middle-class US child to the age of seventeen is estimated at $233,610 (Lino 2017), dwarfing the $8,100 climate impact. And clearly, the value of having a child is worth far more to parents than this monetary cost.

CHAPTER 7. WHY THE GREEN REVOLUTION ISN'T HERE YET

1. **"now economic or . . .":** Lovins, 1976, 83; **Governments around the:** IEA, 2018, 256, used for solar PV and wind.

2. **Ending our reliance:** We will see the costs of smart climate policies in chapter 11. The undiscounted costs from 2015 to 2050 of eliminating fossil fuels is close to $400 trillion in lost GDP using Nordhaus's DICE model.

3. **This is almost:** Smil, 2014; **"Suggesting that . . .":** Hansen, 2011a. He restated this point in 2018 in the *Boston Globe:* "The notion that renewable energies and batteries alone will provide all needed energy is fantastical. It is also a grotesque idea, because of the staggering environmental pollution from mining and material disposal, if all energy was derived from renewables and batteries. Worse, tricking the public to accept the fantasy of 100 percent renewables means that, in reality, fossil fuels reign and climate change grows" (Hansen, 2018).

4. **That is why:** A related problem is that all solar panels produce power *at the same time* (when the sun is shining). This means that the dollar value of the solar power quickly drops, often to zero, because there is an oversupply on clear days around noon. Models show that when California, Texas, or Germany gets 15 percent of its electricity from solar, its average price will drop by half. When California gets 30 percent of its electricity from solar, the electricity price from solar will drop by more than two-thirds (Sivaram and Kann, 2016). This means that solar will find it ever harder to compete. When solar becomes competitive with fossil fuels, meaning that solar panels will earn an unsubsidized profit, adding more solar will make its price drop and render it uncompetitive again (Wanner, 2019). The same logic holds for wind power, where all producers make electricity only when windy.

5. **today the United States:** Caldeira, 2019.

6. 1900–1948 from EIA, 2012, figure 5; 1949–2018 from EIA, 2019c; 2019–2050 prediction to 16 percent in 2050 from reference scenario EIA, 2019a.

7. **almost unchanged from:** EIA, 2017a.

8. 1800–1900 (Fouquet 2009), 1900–1979 (Benichou, 2014; Etemad and Luciani, 1991), 1971–2018 (IEA, 2019g; 2019d). The International Energy Agency has two predictions, Stated Policies and Current Policies for 2040 (IEA, 2019g). The UN has five main scenarios for the twenty-first century, here shown up to 2050 (IIASA, 2018; Riahi et al., 2017).

9. **The International Energy:** IEA, 2019g.

10. **Half a century:** IEA, 2014b; **Since then, with:** World Bank, 2019d.

11. **"Energy is the . . .":** Reuters, 2019; **Indeed, the International:** IEA, 2014a, for sub-Saharan Africa. IEA finds similar results in 2019 (IEA, 2019g), calling it the "Africa Case," with a GDP increase per person of $3,600, but the analysis treats it more as independent growth leading to more energy consumption.

12. **"Dharnai refused to . . .":** Greenpeace, 2014; **The world's media:** Roy, 2014.

13. **Because solar power:** Vaidyanathan, 2015.

14. **When he showed up:** Vaidyanathan, 2015.

15. **"no doubt that . . .":** Bainimarama, 2013.

16. **"meet the resilience . . .":** Hills, Michalena, and Chalvatzis, 2018.

17. **One common anecdote:** Furukawa, 2014; **It also shows:** Aklin et al., 2017; **In Tanzania:** Lee et al., 2016.

18. **Moreover, when asked:** Grimm et al., 2019; **Even when including:** Kashi, 2020.

19. **The Energiewende has:** Shellenberger, 2019; **Electricity costs have:** IEA, 2019a, 128; and IEA, 2019a; **Germans will have:** Shellenberger, 2019.

20. **This massive expenditure:** Shellenberger, 2019.

21. **In the first:** IEA, 2019d.

22. **In the larger:** IEA, 2019d; **In total, solar:** IEA, 2019d; **The problem for:** Norton et al., 2019; **Needless to say:** Norton et al., 2019; Sterman, Siegel, and Rooney-Varga, 2018.

23. **Yet, the cost:** The average of three models shows a 1.03 percent of GDP cost of EU's 20-20-20 climate policy at its most effective, but realistically a 2.19 percent cost (Bohringer, Rutherford, and Tol, 2009). This is about $408 billion; **Indeed, about 20 percent:** EPRS, 2019.

24. **Today, residential:** Thirteen cents for the United States (EIA, 2019b), 23 euro cents, or 25.6 cents for the EU (Eurostat, 2019a); **Over the next:** Panos and Densing, 2019.

25. **"clean our clocks":** Friedman, 2005.

26. **"at least $26 trillion":** Guterres, 2018.

27. **The UN estimates:** IPCC, 2018, 154; **A 2018 Goldman Sachs:** Trivedi, 2018.

28. **If the European Union:** The average estimated cost of the EU 80 percent greenhouse gas emissions reduction by 2050 is a loss of 5.14 percent of GDP as estimated by seven regional models (Knopf et al., 2013). This assumes all policies are perfectly effective. More realistically, the costs will double, as they did for the EU's 20-20-20 climate policy (Bohringer, Rutherford, and Tol, 2009). That leads to a 10.3 percent cost, or €2.514 trillion or $2.8 trillion; **This is more:** Eurostat, 2019b, table gov_10a_exp.

29. **This is why:** UNEP 2019, 3: "The current level of global greenhouse gas emissions is by now almost exactly at the level of emissions projected for 2020 under the business-as-usual, or no-policy, scenarios used in the Emissions Gap Reports, which are based on the assumption that no new climate policies are put into place from 2005 onwards. In other words, essentially there has been no real change in the global emissions pathway in the last decade."

CHAPTER 8. WHY THE PARIS AGREEMENT IS FAILING

1. **It was hailed:** UNFCCC, 2018b; **French president:** Vidal et al., 2015; **"This is a historic . . .":** Vidal et al., 2015; **Al Gore saw:** Gore, 2015.

2. **In promises made:** USNDC, 2016; **Work from the:** Fawcett et al., 2014.

3. **The EU promised:** EUNDC, 2016; **There is no:** Knopf et al., 2013.

4. **It set a 2030:** China NDC, 2016; **Its results suggest:** Calvin et al., 2012; Calvin, Fawcett, and Kejun, 2012.

5. **Mexico has enacted:** Mexico NDC, 2016; **Although Mexico:** Veysey et al., 2016.

6. **For instance, back:** It also made a number of other promises, but this was the most important one; **Stanford's Energy Modeling:** Bohringer, Rutherford, and Tol,

2009. The "forever" part means that the EU would now be on a lower growth path, so even if growth rates without new climate policies were similar from 2021 and onward, it would always be about 1 percent behind; **In total, the:** Bohringer, Rutherford, and Tol, 2009. Notice, the EU itself, using one very optimistic model, estimated the total cost at an unbelievable 0.5 percent of GDP.

7. **Whereas the cheapest:** Young and Bistline, 2018.

8. **Thus, without:** If anything, this is likely a low-ball estimate. A new study indicates that the annual global cost of the Paris Agreement could be $5.4 trillion (J. Li, Hamdi-Cherif, and Cassen, 2017).

9. **No matter which:** SIPRI, 2019; **Every year, the:** Gomes, 2010, 47; **And to put:** UNEP, 2014, 435; **It is also:** UNAIDS, 2019, 174.

10. **The UN estimates:** IPCC, 2013b, 27. With the transient climate response to cumulative carbon emissions likely in the range of 0.2°C to 0.7°C per 1,000 Gt carbon dioxide (0.8°C–2.5°C per 1,000 GtC, IPCC, 2013b, 16–17), and 0.45°C being perhaps the most realistic (Kriegler Elmar et al., 2018, 3; Matthews, Solomon, and Pierrehumbert, 2012, 4369).

11. **This estimates:** UNFCCC, 2015; **According to the:** A reduction of maximally 63.8 Gt carbon dioxide will reduce global temperatures about 0.029°C and certainly less than 0.045°C (similar to Lomborg, 2016).

12. From Lomborg, 2020.

13. **A 2017 landmark:** Victor et al., 2017.

14. **The story is:** Victor et al., 2017.

15. **In contrast, the:** CAT India, 2016.

16. **A 2018 study:** Nachmany and Mangan, 2018.

17. **Translated into an:** The UN relation of 0.8°F per 1,000 Gt carbon dioxide reduced gives 0.43°F (0.24°C), with two runs showing 0.36°F (0.2°C) (MIT, 2015) and 0.3°F (0.17°C) (Lomborg, 2016).

18. **The global warming:** CAT, 2018.

19. **Indeed, it turns out:** Lomborg, 2020.

20. **More than sixty:** Sengupta and Popovich, 2019.

21. **In 2007, Prime Minister:** ECOS, 2007; **The latest official:** Ministry for the Environment (New Zealand), 2019a, shows 81.4 Mt versus 81.0 Mt in 2007 (Ministry for the Environment [New Zealand], 2019b). **New Zealand is:** Using 1990 as 100 percent, her promise was 0 percent of the 1990 level, but the actual emissions in 2020 will likely be 123 percent of 1990 emissions; **Legislation aimed at:** Tidman, 2019.

22. **For a small:** NZ Treasury, 2019, table 4. All of the dollar amounts for New Zealand costs are converted into US dollars.

23. **Getting all the:** NZIER, 2018, 16. Average of three ZNE scenarios, top table 11 (NZD 87.3 billion). All the dollar amounts are in USD; **That is more:** NZ Treasury, 2019, table 4. All Core Crown expenses, and given that the NZIER $87.3 billion is in 2015 dollars, the cost would be even higher in 2019 dollars.

24. **To achieve their:** A $1,500 carbon dioxide tax is equal to $8.33 per gallon ($3.5/liter).

25. **If the policies:** This is using the report's own GDP growth, linearly scaling in the cost of 16 percent from zero in 2020 to 16 percent in 2050, and then staying at 16

percent for rest of the century. The per person cost uses the average population of New Zealand over the century, which the UN expects to be 5.6 million people (UN-DESA, 2019).

26. **As a back-of-the-envelope:** US GDP in 2050 according to UN SSP2 (IIASA, 2018) is 27 trillion PPP$ 2010; adjusted to 2018 dollars, 16 percent is $5 trillion. **That is higher:** Amadeo, 2019.

27. **If we assume:** Total reduction is about 5.4 Gt carbon dioxide over the century; using 0.8°F per 1,000 Gt carbon dioxide means a 0.004°F (0.0025°C) temperature reduction; **Given the expected:** As the temperature rises by 0.0675°F per year in the decade 2090–2100 for the SSP2 (IIASA, 2018), this means that the 0.004°F temperature reduction will push forward the temperature rise seen on January 1, 2100, by 23.5 days.

CHAPTER 9. PICK A PATH: WHICH FUTURE IS BEST?

1. **By 2030, the:** As we saw in the cost estimate for the Paris Agreement in chapter 8.

2. **This sounds like:** The SSP work was outlined from 2014 (O'Neill et al., 2014, and van Vuuren et al., 2014, and published in a special issue in 2017 (Riahi et al., 2017); **They examined five:** (Riahi et al., 2017).

3. **That prediction is:** Bolt et al., 2018; Maddison, 2006.

4. All amounts are in $2011 PPP. To the right, the increase in GDP per person in 2100 compared to GDP per person in 2020. In 2100, the average person in the fossil fuel scenario will see a GDP per person worth 1,040 percent of the average person in 2020 (IIASA, 2018; Riahi et al., 2017).

5. **There is slow:** Fricko et al., 2017; **In a 2018 survey:** Christensen, Gillingham, and Nordhaus, 2018.

6. **The final pathway:** Jiang, 2014.

7. **That may sound:** Christensen, Gillingham, and Nordhaus, 2018.

8. **Both of these:** Rao et al., 2019; **over the next:** This is an average over the years 2020–2050., some years more, some years less; **But especially the:** Riahi et al., 2017 figure 2D; **By 2100 under:** Lomborg, 2020.

9. **In the bottom:** KC and Lutz, 2017, 189; **Illiteracy, which is:** Riahi et al., 2017, 158; **Globally, life expectancy:** Lutz, Butz, and KC, 2014, 669.

10. Global GDP per person for the green and the fossil fuel scenarios (IIASA, 2018). The dashed lines show GDP per person without taking climate into account. The full lines show GDP per person minus climate damages as Nordhaus has estimated in figure 17. Thus, the cooler sustainable scenario results in a temperature rise of 5.83°F in 2100. This implies a 2.5 percent reduction of nominal GDP per person to estimate actual welfare. The warmer fossil-fuel-driven scenario will reach a temperature of 8.75°F in 2100, which will result in a more damaging 5.7 percent reduction from nominal GDP per person.

11. **In late 2018:** *Guardian*, 2018; **They and global:** Institutet för nervväxtstudier, 2018.

12. **Researchers have looked:** Keeney, 1990; Lutter and Morrall, 1994; Lutter, Morrall, and Viscusi, 1999; Broughel and Viscusi, 2017; Hahn, Lutter, and Viscusi, 2000;

Using the estimates: Lomborg, 2020. Since all humans die, we don't actually avoid these three million deaths, but we avoid three million people *dying prematurely.*

CHAPTER 10. HOW CLIMATE POLICY HURTS THE POOR

1. **"The Rich Pollute . . .":** *Economist,* 2017.
2. **"Cut Carbon . . .":** Elliott and Seager, 2007.
3. **Typically, whatever the:** Aronoff, 2019.
4. **Let's look at:** NDRRMC, 2014. **Built on flat:** Athawes, 2018; **In 1912, a:** Galvin, 2014; *Washington Herald,* 1912. Indicating that "half the population" was lost, of a Tacloban population of twelve thousand.
5. **According to the** *Guardian***:** Vidal, 2014; **The diplomat vowed:** Climate Home News, 2013.
6. **One 2016 study shows:** Twenty-seven percent, according to Bakkensen and Mendelsohn, 2016.
7. **Nonetheless, investment in:** Assuming that GDP per capita and vulnerable infrastructure grow at same rate.
8. **It also helps:** Bakkensen and Mendelsohn, 2016.
9. **But a comprehensive:** Hasegawa et al., 2018.
10. **A 2019 study:** Campagnolo and Davide, 2019.
11. Data from IIASA, 2018. Climate damages estimated by Nordhaus's RICE model (W. Nordhaus, 2013).
12. **Using a regionally:** W. Nordhaus, 2010; 2013.
13. **Campaigners used to:** Connor, 2014; **We don't often:** Roser and Ritchie, 2019b.
14. **A study shows:** Tol and Dowlatabadi, 2001.
15. **Yet higher costs:** Tol and Dowlatabadi, 2001.
16. **Twenty years ago:** Guo, Song, and Buhain, 2015, 716; **The European Union:** European Union, 2003, Article 3 (b), ii; **Developing nations, even:** Monbiot, 2007.
17. **This movement originally:** WWF, 2007; NDRC, 2008; **The charity ActionAid:** ActionAid, 2012.
18. **The huge growth:** Chakrabortty, 2008; **After food prices:** Nebehay, 2008; **The World Bank:** Ivanic, Martin, and Zaman, 2011.
19. **Many environmental groups:** Witness the difference in tone from the NDRC just from 2008 to 2009 (NDRC, 2008; 2009); *Guardian* **columnist and:** Monbiot, 2007.
20. **One 2019 study:** Chambers, Collins, and Krause, 2019.
21. **The International Energy Agency:** IEA, 2017, 25.
22. **Cold homes are:** Kahouli, 2020; **The study estimates:** Chirakijja, Jayachandran, and Ong, 2019.
23. **People will be:** The actual increase in deaths will likely be larger, since this study only looked at natural gas use, whereas a comprehensive climate policy also will affect the many other ways to heat one's home.
24. **In the United Kingdom:** Ofgem, 2018; **Ever more stringent:** Department for Business, Energy & Industrial Strategy, 2019. Using 2006–2016; and Office for National Statistics, 2017, latest UK numbers for individuals. **Not surprisingly:** NEED, 2019.

25. **One poll in:** *Belfast Telegraph,* 2014.

26. The amount measures both aid spent on mitigation and adaptation, and both as a principle and a significant objective (Hicks, 2008, 37; OECD-DAC, 2019; OECD, 2019).

27. **In general, studies:** Lomborg, 2018.

28. **Indeed, they will:** IEA, 2017, 53; **That is half:** IGS, 2019; **It won't even:** Institute for Health Metrics and Evaluation, 2019.

29. **It means households:** Khandker, 2012.

30. **In one 2016 study:** Gunatilake, Roland-Holst, and Larsen, 2016.

31. **Denying Bangladesh this:** Of course, the actual costs for Bangladesh are much higher, because it also has to pay for the power plants. For Bangladesh, the benefits outweigh the costs by about twenty-five times; **And Bangladesh is:** World Bank, 2017.

32. **We have to:** Hance, 2017.

33. **Today there are:** Roser and Ortiz-Ospina, 2019a; **It turns out:** Ortiz-Ospina, 2017. They estimate this with eight hundred million poor; the cost would be even lower at 650 million poor.

CHAPTER 11. CARBON TAX: THE MARKET-BASED SOLUTION

1. **Indeed, it doesn't:** W. Nordhaus, 2013, 6–7.

2. **Unfortunately, another:** This is true of any tax, e.g., Romer and Romer, 2010; **It forces people:** Tol, 2019, 32.

3. **Using Nordhaus's model:** We get the $140 trillion by taking the full cost across five centuries and expressing it as if it all had to be paid today. This means that if we put $140 trillion into an account today with an interest of about 4 percent per year, we would have exactly enough to pay out for all the net damages from climate change for the next five hundred years.

4. W. Nordhaus, 2018.

5. **A carbon tax:** Tol, 2019, 35–39.

6. W. Nordhaus, 2018.

7. **The climate-economic model:** Notice that Nordhaus's model actually includes the slight cost of policies already enacted.

8. **This will mean:** Notice that this policy cost is also expressed as if we were to pay all of it today, to be comparable with the climate cost discussed above.

9. W. Nordhaus, 2018.

10. **Notice, we don't:** As Nordhaus puts it: "A limit of 2°C appears to be infeasible with reasonably accessible technologies even with very ambitious abatement strategies" (W. Nordhaus, 2018, 334).

11. **In Nobel laureate:** W. Nordhaus, 2018.

12. **Even though the:** W. Nordhaus, 2018.

13. W. Nordhaus, 2018.

14. **Look at France:** Willsher, 2018. Thirteen cents per gallon, four cents per liter of gasoline, or three euro cents per liter.

15. **It turns out:** Akimoto, Sano, and Tehrani, 2017.

16. **As we saw:** Bohringer, Rutherford, and Tol, 2009; **Similarly, much climate:** Young and Bistline, 2018.

17. Lomborg, 2020.

18. Lomborg, 2020.

19. The total value of the next five hundred years of GDP is expressed in net present value. It is $4,629 trillion, which, if invested today at realistic interest rates, would be able to exactly pay out the expected GDP each year over the next five hundred years.

20. **Yet, the impact:** If we do not manage global buy-in for carbon taxes, the impacts will be even smaller, although the costs will likely similarly decrease. Carbon taxes of the same magnitude would probably still be effective for individual nations, although some carbon emissions would likely migrate to nations with low or no carbon taxes.

CHAPTER 12. INNOVATION: WHAT IS NEEDED MOST

1. **From the 1700s:** Throughout most of the 1700s, whale oil was only for the very rich, but by 1850 it had a 65.5 percent market share of oil and gas illuminants (Kaiser 2013, 9). Kerosene, just one year after the first petroleum drilling in Pennsylvania, had a 20 percent market share in 1860; **At its peak:** Dolin, 2008, 242; L. E. Davis, Gallman, and Hutchins, 1988.

2. **Before the turn:** Stephen Davies, 2004.

3. **"the battle to . . .":** Ehrlich, 1968, 11.

4. **The belief was:** Ehrlich, 1968, 160; **Conventional wisdom:** Ehrlich, 1968, 40–41.

5. **As a result, grain:** Encyclopedia Britannica, 2020.

6. **By 1980, India:** FAO, 2019; Roser, Ritchie, and Ortiz-Ospina, 2019; **Calories available:** Roser and Ritchie, 2019a; **Today, India is:** Statista, 2019.

7. **We would have:** Bailey, 2013.

8. **The International Energy:** IEA, 2019g; **The cheapness of:** IEA, 2019g.

9. **Fracking has reduced:** Chirakijja, Jayachandran, and Ong, 2019; **It has also dramatically:** Melek, Plante, and Yücel, 2019.

10. **It has also had:** Loomis and Haefele, 2017; **The biggest study:** This cost appears to ignore that fracking will also lead to *lower* particulate air pollution where the gas is burned instead of coal (often not in the United States, where fracking took place, or where the study focuses). This estimate is a *benefit* of about $17 billion per year, cancelling about three-quarters of the disbenefits from fracking (Johnsen, LaRiviere, and Wolff, 2019).

11. **The technology of:** Golden and Wiseman, 2015; **The fracking innovation:** Golden and Wiseman, 2015; **Crucially, gas emits:** Fracking also emits more methane (which is a greenhouse gas), and some had speculated that the extra emissions of methane could outweigh the lower carbon dioxide emissions. All the major studies, including from the US Environmental Protection Agency, show that this is extremely unlikely, and that after taking methane into account, switching from coal to natural gas in the power sector produces climate benefits over all time frames (Raimi and Aldana, 2018); **This is the:** EIA, 2018; and using data from the Global Carbon Project,

2019b, showing a reduction in carbon dioxide emissions of 511 Mt CO_2, with the UK reducing 166, Italy reducing 129, and Ukraine reducing 101 Mt CO_2.

12. **In fact, in:** BP, 2019; **If China switched:** Given that air pollution from coal in China is massive, more fracking for gas would likely also lead to cleaner air for most people in most places, although worse air pollution at fracking sites.

13. **We worked with:** Lomborg, 2010.

14. **Since then, my:** We even managed to get the cover story on the *Guardian* in 2010, calling for the $100 billion R&D fund (August 10, 2010); **The most promising:** Cama, 2015.

15. **International Energy Agency:** IEA, 2019c.

16. Green investment is taken as all R&D investment minus fossil fuel investment (so energy efficiency, renewables, nuclear, hydrogen, and fuel cells, other power and storage technologies, other cross-cutting technologies/research and unallocated) (IEA, 2019c). GDP from the World Bank, 2019e, adjusted to 2018 dollars with BEA, 2019.

17. **Globally, private companies:** IEA, 2019f, 160; **As a percentage:** IEA, 2019f, 160.

18. **Globally, in 2020:** IEA, 2018, 256.

19. **Before the Chicago World's Fair:** Walter, 1992; **There were a:** Walter, 1992, 117; Walter, 1992, 59; Walter, 1992, 67; Walter, 1992, 26.

20. **E-mail was foreshadowed:** Walter, 1992, 66; **Air travel would:** Walter, 1992, 187; **Multiple thinkers expected:** Walter, 1992, 213.

21. **One considered how:** Walter, 1992, 66; **Others realized that:** Walter, 1992, 60, 144. Actually, since then electricity became forty to sixty times cheaper and indeed powered the United States and the world. The retail price of a kilowatt-hour in the 1880s was 28¢ (Schobert, 2002, 188), or in today's money $7.60 (CPI, 2019). In 1902, the price was 16.2¢ (Census, 1975, vol. 2, S116) or $5. In 2018, the average retail price was 12.87¢ (EIA, 2019a).

22. **But these make:** Gür, 2018, 2699.

23. **My personal favorite:** Rathi, 2018.

24. **If we exclude:** BNEF, 2019, estimates 7 GWh today, with the world using 0.84 GWh per second (IEA, 2019g); **Headlines consistently tell:** John, 2019; **The International Energy Agency:** IEA, 2019g, 253; **Yet, this will:** The IEA doesn't express its 330 GW capacity in energy, but using the same conversion from BNEF, 2019, it is equivalent to 859 GWh, which at the higher electricity consumption of 1.3 GWh per second is almost eleven minutes. The much higher BNEF estimate of 2,850 GWh is equivalent to thirty-six minutes in 2040; **While utility-scale solar:** Lazard, 2019a, 2; Lazard, 2019b, 4.

25. **It has promised:** Sixty-two GW in 2018 (IEA, 2019e, 31) to 620 GW in 2040 (IEA, 2019g, 295); **Notice, this won't:** IEA, 2019g; **A lot of:** IEA, 2019g, 294; **Partly as a:** IEA, 2019g, 296.

26. **The International Energy Agency:** IEA, 2019g, 296–97. Seeing a cut of 320 Mt carbon dioxide, or a little less than one percent of global fossil fuel emissions.

27. **Nuclear energy doesn't:** It actually emits greenhouse gasses when built and decommissioned, but on a life cycle basis this is a tiny contribution (IPCC, 2014b, 539); **Under normal operating:** Because coal has trace amounts of radioactive materials, released during combustion (Chiras, 1998, 266); **Although we think:** Markandya and Wilkinson, 2007.

28. **The reason why:** Lazard, 2019a. And decommissioning of nuclear power plants can cost ten times their construction costs (Cunningham and Cunningham, 2017, 441); **Finland's newest nuclear:** Deign, 2019; **It may open in 2021:** Edwardes-Evans, 2019; **It may open in 2022:** AFP, 2019.

29. **A 2017 survey:** Lang, 2017; **In France and:** Grubler, 2010, 5182–83, and Berthélemy and Escobar Rangel, 2015; **One 2017 study suggests:** Lang, 2017.

30. **Philanthropist Bill Gates:** TerraPower, 2019; Reuters, 2017; **China is demonstrating:** Z. Zhang, 2019; **Others are suggesting:** N. Johnson, 2018.

31. **One 2019 study:** Lazard, 2019a, 2, finds nuclear at 11.8¢–19.2¢ per kWh, whereas EIRP, 2017, finds average costs of 6¢, with the best case at 3.6¢, below gas at 4.4¢.

32. **A third area:** Sanz-Pérez et al., 2016; Pires, 2019; Roger A. Pielke, 2009.

33. **In 2007, entrepreneur:** Branson, 2019.

34. **In 2011, an:** APS, 2011; **A 2017 overview:** Ishimoto et al., 2017; **We should take:** Keith et al., 2018.

35. **If we can:** This is at the UN's middle-of-the-road scenario with 79 billion tons of carbon-dioxide-equivalent emissions, $5 per ton (or $395 billion in total cost) of a $231 trillion economy (IIASA, 2018).

36. **Because the algae:** Herper, 2017.

37. **Fixing global warming:** An annual cost of $192–$408 billion, or 1.03–2.19 percent of GDP (Bohringer, Rutherford, and Tol, 2009).

CHAPTER 13. ADAPTATION: SIMPLE BUT EFFECTIVE

1. **Not only do:** Formetta and Feyen, 2019.

2. **At its simplest:** Auffhammer and Mansur, 2014; **On the other hand:** Berrittella et al., 2006.

3. **In contrast, adaptation:** Kongsager, 2018, 8; **Indeed, most adaptation:** Fankhauser, 2017, 215.

4. **In the rich world:** Agrawala et al., 2011, 29.

5. **Where it is:** Seo and Mendelsohn, 2008.

6. **Across the world:** M. Chen et al., 2018; Gezie, 2019; Bakhsh and Kamran, 2019.

7. **In Ethiopia, a:** Di Falco, Veronesi, and Yesuf, 2011.

8. **Yet weirdly, for:** Kongsager, 2018.

9. **In fact, sea:** Thirty-one cm, or 1.01 foot, from an average of 1850–1870 up to 2010 (Jevrejeva et al., 2014).

10. **One of the:** Lincke and Hinkel, 2018; Markanday, Galarraga, and Markandya, 2019; Hinkel et al., 2014; **The study shows:** Hinkel et al., 2014.

11. **The total cost:** Lincke and Hinkel, 2018, table S1.

12. **"Coastal defense" in:** Markanday, Galarraga, and Markandya, 2019, 19.

13. **The benefits of:** Global Commission on Adaptation, 2019, 14, 31.

14. **Holland has shown:** EEA, 2018; Rijke et al., 2012.

15. **If there were:** Numbers for the United States and Canada (Lim et al., 2018, figures 2a and 3a); **The National Institute:** NIBS, 2018; **For the EU:** Rojas, Feyen, and Watkiss, 2013.

16. **It is in fact:** Satija, Collier, and Shaw, 2016.

17. **To claim that:** Berke, 2017; **Since the 1980s:** *Houston Chronicle,* 2017; **One 2018 study:** Muñoz et al., 2018.

18. **In 2019, Houston:** Anchondo, 2019; **A 2019 study:** Davlasheridze et al., 2019.

19. **In one area:** Baddour, 2017.

20. **Encouraging homeowners:** Markanday, Galarraga, and Markandya, 2019.

21. **Although some 90 percent:** Hossain, 2018.

22. **In 1991, at:** Bern et al., 1993; **Since 1991, a:** Paul, 2009, and Haque et al., 2012; **In the 2010s:** Centre for Research on the Epidemiology of Disasters, 2019.

23. **Overall, it has:** Global Commission on Adaptation, 2019, 49; **For river flooding:** Markanday, Galarraga, and Markandya, 2019; **For heavy precipitation:** Markanday, Galarraga, and Markandya, 2019.

24. **Building codes matter:** Flavelle, 2018; **If this code:** NIBS, 2018, 2, 58.

25. **The city suffered:** Montana State University, 2019.

26. **A fire expert:** Kaufman, 2018.

27. **Cities are increasingly:** IIASA, 2018.

28. **For example, during:** Kenward et al., 2014.

29. **In the aftermath:** Biello, 2014; **It has already:** City of New York, 2019; **Los Angeles has:** McPhate, 2017; **Theoretical models show:** Georgescu et al., 2014.

30. **In London, the:** Wilby and Perry, 2006, 92.

31. **If we increase:** Greater London Authority, 2006, 11; **And analysis from:** Greater London Authority 2006, 12.

32. **One 2017 study:** Estrada, Botzen, and Tol, 2017. **Unfortunately, green roofs:** Estrada, Botzen, and Tol, 2017. Another study finds "green measures" deliver about two dollars back on the dollar (Markanday, Galarraga, and Markandya, 2019).

33. **Those with swimming:** Adler et al., 2010.

34. **In Chicago, such:** Adler et al., 2010, 33; **France has taken:** P. Ford, 2019, 20.

35. **After a 2010:** Kaur, 2017; **A similar heat wave:** Ahmedabad Municipal Corporation, 2016.

36. **In 2019, Mayor:** Goldman and Flavelle, 2019.

37. **Paradise is a working-class:** Kaufman, 2018.

CHAPTER 14. GEOENGINEERING: A BACKUP PLAN

1. **The eruption injected:** Proctor et al., 2018.

2. **Scientists have come:** Baskin, 2019.

3. **Some fear that:** Bradford, 2017; **Others worry that:** Meyer, 2018.

4. **One of the cheapest:** NRC, 2015, 101ff.

5. **This fleet would:** Mims, 2009.

6. **This is a tantalizing:** The $140 trillion discussed earlier is the cost over the next five centuries.

7. **If changing the:** J. L. Reynolds and Wagner, 2019.

8. **What if they're right?:** Catastrophic impacts was one of the reasons Nordhaus added 25 percent to the cost of climate change, and it is included in the models we

have used to find the best carbon tax; **Potential pitfalls are:** Geoengineering wouldn't be straightforward to implement, because not all countries would actually have the same interests. Imagine if humanity actually had the power to control the global thermostat. In whose hands should this power be held? Tropical countries might very well want to turn the temperature down, but colder Russia, Canada, and Norway might want it cranked up. Bear in mind that for some nations like the United Kingdom and Poland, short-term global warming means net benefits, especially for agriculture. Reducing the average global temperature isn't in every nation's short-term interest, making negotiations even more delicate. It is hard to imagine leaders actually agreeing on the "right" temperature for the world. In that sense, the only realistic goal will probably be to return to what the temperature "used" to be, because this is a focal point for the world (providing the only obvious temperature to agree on) and also what most of the world has already adapted to.

9. **A 2019 survey:** Dannenberg and Zitzelsberger, 2019.

10. **The key arguments:** Geoengineering Monitor, 2020.

11. **If we reduce:** Proctor et al., 2018.

12. **Most people experience:** Irvine et al., 2019.

13. **The same reasoning:** Corner, 2014; Hamilton, 2015.

14. **Even in the face:** Kintisch, 2017; **And in the wake:** Bawden, 2016.

15. **They found that:** Bickel and Lane, 2009; **One of the:** Kunzig, 2008.

16. **This would be:** Lane et al., 2009.

CHAPTER 15.
PROSPERITY: THE OTHER CLIMATE POLICY WE NEED

1. **Both countries are:** Bos and Zwaneveld, 2017, 12; Dasgupta et al., 2011, 168.

2. **the Delta Works:** Aerts, 2009, 40; **Since 1953, there:** EM-DAT, 2020.

3. **In contrast, Bangladesh:** Notice that "normal" flooding is actually beneficial in Bangladesh, as it supplies water for crop production, recharges the groundwater table, replenishes soil fertility, and supports fish production (Banerjee, 2010). The problems come from extreme flooding, which happens every four to five years (Dasgupta et al., 2011); **In 2019, flooding:** AlJazeera, 2019; **In the first:** EM-DAT, 2020; **Every year, flooding:** Ferdous et al., 2019.

4. **As Bangladesh gets:** This is a global phenomenon (Jongman et al., 2015); **Flood defenses for:** Dasgupta et al., 2011,

5. IIASA, 2018; W. Nordhaus, 2013.

6. **Today, Bangladesh spends:** IEA, 2020.

7. **The Netherlands is:** The World Bank, 2019e, shows that Bangladesh lost forests on 0.3 percent of its land area in this century, while the Netherlands planted extra forest on 0.5 percent of its land area; **While the area:** Schuerch et al., 2018; **In total, richer:** Schuerch et al., 2018.

8. **Back in 1992:** Schelling, 1992.

9. **One of the clearest:** Letta, Montalbano, and Tol, 2018.

10. **Indeed, a 2012 study:** Anthoff and Tol, 2012.

CHAPTER 16. CONCLUSION:
HOW TO MAKE THE WORLD A BETTER PLACE

1. **Figure 16.1 is:** Lomborg, 2013. To ensure comparability, all analyses estimated the cost of not solving the problem, so the cost of malnutrition is calculated by establishing how much richer the world would have been every year if everyone were well fed, hence more productive and healthier. Likewise for health (how much richer the world would be if easily curable diseases were healed) or for education (if much of the world had not been illiterate for generations). To ensure the costs were comparable across long periods of time, only the "permanent" part of the problem was investigated (education has many challenges, but the estimate only shows the cost of illiteracy), so these are definitely underestimates. Bearing this in mind, the figure shows that the world has moved dramatically toward less costly problems.

2. **Second, the figure:** The original analysis included the cost of global warming from 1900 to 2050, but as I outlined in chapter 5, the costs were so small (and were even slightly positive in the past) that they wouldn't have been visible in this graph. It also included the cost of biodiversity and water and sanitation, both of which were also very small, and not shown here. Instead, the graph shows the cost of climate change with an optimal policy from 2015 to 2100, as calculated by William Nordhaus. It reaches a cost of 2.86 percent of GDP in 2100. The equivalent cost if the world were to do *no* climate policy whatsoever would be 3.98 percent by 2100.

3. Based on estimates in Lomborg, 2013, the estimate for each problem shows how much richer the world could have been, had the problem been addressed (so the cost of all the problems can add to more than 100 percent). For reference, the cost in percentage of global GDP of Nordhaus's optimal 3.5°C scenario 2015–2100 is shown.

4. **Half the world's:** M. Smith, 2019.

5. **He was a:** Hall, 2009; **"Sometime in . . .":** Retro Report, 2019, 5:40. "Nonetheless, Ehrlich continues to claim that civilizational collapse is just a couple of decades away" (Carrington 2018); **Other influential academics:** Environmental Action, 1970, 25; *New York Times,* 1970; *Life* **magazine:** *Life,* 1970.

6. **Ehrlich predicted that:** Predicted "the death rate will increase until at least 100–200 million people per year will be starving to death during the next ten years" (Collier 1970, 293); **He was off:** About 3.3 million people starved to death in the 1970s, more than half from Vietnam's invasion of Cambodia in 1979 (Hasell and Roser, 2017); **He claimed that:** Editors of Ramparts, 1970, 7; **Los Angelinos didn't:** Parrish and Stockwell, 2015.

7. **"the only possible . . .":** Editors of Ramparts, 1970, 9; **Some researchers started:** Ehrlich, 1968, 135; **It gave urgency:** R. J. Williams, 2014, 485; **In just one:** Biswas, 2014.

8. **Believing that continued:** Ehrlich, 1968, 78–80.

9. **In large part:** Carson, 1962; **But even at:** Carson, 1962, 227; WHO, 2019; **Even the United Nations:** *Canberra Times,* 1977; Doll and Peto, 1981, 1256.

10. **The resulting nationwide:** This was not least true for EPA in its first decades (Colborn, Dumanoski, and Myers, 1996, 202); **The total cost:** Graham, 1995, 1; **A research team:** Tengs et al., 1995; Tengs, 1996; Tengs and Graham, 1996; **Benzene emission controls:** Graham, 1995.

11. **Across 185 life-saving:** Tengs and Graham, 1996; **However, if the:** Tengs and Graham, 1996; **Put bluntly, the:** And this number is undoubtedly much too low, since the Harvard study could only assess regulations already evaluated for costs and benefits. A different way to see this imbalance is that in the first decades of EPA, its typical lifesaving regulations delivered benefits worth just 4¢ on each dollar of public expenditure (Tengs et al., 1995), findings generally reconfirmed in Morrall III, 2003. The cost-benefit estimate is based on the median cost for the EPA and other agencies, using the EPA's own estimate of the value of saving one life in one year—$324,000 ($293K in 1990$) (EPA, 1997, ES-6)—updated to 1993$ with CPI 2019. Other federal agencies, not affected by toxin fear, wasted far fewer resources; typically they saw benefits of three to fourteen times their costs.

12. **"every city ...":** Eisenhower, 1961.

13. **Statistics on Soviet:** Trachtenberg, 2018.

14. **In just the:** Woerdman, Couwenberg, and Nentjes, 2009; Hintermann, 2016; **Energy companies' lobbying:** Brulle, 2018, 295.

15. **"If implemented ...":** Morgan, 2002. The memo can be found in Palmissano, 1997.

16. **Enron boasted of:** Palmissano, 1997.

17. For instance, doubling renewable energy will produce economic benefits of better energy access, and environmental benefits of lower CO_2 emissions; for each dollar spent, the total social, environmental, and economic benefits are worth 80¢. Universal access to contraception will reduce maternal and child mortality, increase investment in child rearing, and lead to higher economic growth; for each dollar spent, the total benefits are worth $120. All from Lomborg, 2018.

18. **My think tank:** Copenhagen Consensus Center, 2019.

19. **Globally, freer trade:** Anderson, 2018.

20. **Every dollar spent:** Horton and Hoddinott, 2018; **This research was:** Department for International Development et al., 2013.

21. **For about $6 billion:** Vassall, 2018; **"best investment ...":** Gates, 2019.

22. **There are 214 million:** WHO, 2018; **Each dollar spent:** Kohler and Behrman, 2018; **This research recently:** Department for International Development and Alok Sharma, 2019.

BIBLIOGRAPHY

AAP-Reuter. 1982. "Environment Enthusiasm 'Less.'" 1982. https://trove.nla.gov.au /newspaper/page/13937663.

Achenbach, Joel. 2018. "Extreme Weather in 2018 Was a Raging, Howling Signal of Climate Change." 2018. https://www.washingtonpost.com/science/2018/12/31 /extreme-weather-was-raging-howling-signal-climate-change/.

ActionAid. 2012. "Biofuels: Fueling Hunger? Threats of Another Global Food Crisis." 2012. https://www.actionaid.org.uk/sites/default/files/publications/biofuels _fuelling_hunger.pdf.

Adler, Michael, Samantha Harris, Megan Krey, Loni Plocinsky, and Jeanette Rebecchi. 2010. "Preparing for Heat Waves in Boston—A Cool Way to Attack Global Warming." 2010. https://www.cityofboston.gov/images_documents/Preparing%20for %20Heat%20Waves%20in%20Boston_tcm3-31986.pdf.

Aerts, J. C. J. H. 2009. "Adaptation Cost in the Netherlands: Climate Change and Flood Risk Management." *Climate Research Netherlands—Research Highlights* (2009): 34–44. http://www.climatechangesspatialplanning.nl/media/default.aspx/emma /org/10354094/34-44+Adaptation+cost+in+the+Netherlands+-+Climate+Change +and+flood+risk+management.pdf.

AFP. 2019. "French Nuclear Power Plant Is Seven Years Late and Costs Have Tripled." October 28, 2019. https://www.thelocal.fr/20191028/french-nuclear-power-plant -is-seven-years-late-and-costs-have-tripled.

Agrawala, Shardul, Maëlis Carraro, Nicholas Kingsmill, Elisa Lanzi, Michael Mullan, and Guillaume Prudent-Richard. 2011. "Private Sector Engagement in Adaptation to Climate Change: Approaches to Managing Climate Risks." November 2011. https://doi.org/10.1787/5kg221jkf1g7-en.

Ahmedabad Municipal Corporation. 2016. "Ahmedabad Heat Action Plan 2016." 2016. https://www.nrdc.org/sites/default/files/ahmedabad-heat-action-plan-2016.pdf.

Akimoto, Keigo, Fuminori Sano, and Bianka Shoai Tehrani. 2017. "The Analyses on the Economic Costs for Achieving the Nationally Determined Contributions and the Expected Global Emission Pathways." *Evolutionary and Institutional Economics Review* 14, no. 1 (2017): 193–206. https://doi.org/10.1007/s40844-016-0049-y.

Aklin, Michaël, Patrick Bayer, S. P. Harish, and Johannes Urpelainen. 2017. "Does Basic Energy Access Generate Socioeconomic Benefits? A Field Experiment with Off-Grid Solar Power in India." *Science Advances* 3, no. 5 (2017). https://doi.org/10.1126 /sciadv.1602153.

AlJazeera. 2019. "Worst Floods in Years 'Submerge' Bangladesh Villages." 2019. https:// www.aljazeera.com/news/2019/07/worst-floods-years-submerge-bangladesh -villages-190719083053518.html.

Allen, Myles. 2019. "Why Protesters Should Be Wary of '12 Years to Climate Breakdown' Rhetoric." 2019. https://theconversation.com/why-protesters-should-be-wary-of-12-years-to-climate-breakdown-rhetoric-115489.

Amadeo, Kimberly. 2019. "FY 2019 Federal Budget: Trump's Budget Request." 2019. https://www.thebalance.com/fy-2019-federal-budget-summary-of-revenue-and-spending-4589082.

Anchondo, Carlos. 2019. "Legislation with $1.7 Billion for Flood Control and Mitigation Projects Goes to Governor." 2019. https://www.texastribune.org/2019/05/26/lawmakers-approve-bill-to-help-fund-floor-control-projects-in-texas/.

Andela, N., D. C. Morton, L. Giglio, Y. Chen, G. R. van der Werf, P. S. Kasibhatla, R. S. DeFries, et al. 2017. "A Human-Driven Decline in Global Burned Area." *Science* 356, no. 6345 (2017): 1356–62. https://doi.org/10.1126/science.aal4108.

Anderson, Kym. 2018. "Benefits and Costs of the Trade Targets for the Post-2015 Development Agenda." In *Prioritizing Development*, edited by Bjorn Lomborg. Cambridge, UK: Cambridge University Press, 2018, 192–215. https://www.copenhagen.com.

Annon. 1991. "What's Happening to the World's Weather?" *Earth Island Journal* 6, no. 3 (1991): 38.

Anthoff, David, and Richard S. J. Tol. 2012. "Schelling's Conjecture on Climate and Development: A Test." In *Climate Change and Common Sense: Essays in Honour of Tom Schelling*. Oxford: Oxford University Press. https://www.oxfordscholarship.com/view/10.1093/acprof:oso/9780199692873.001.0001/acprof-9780199692873-chapter-14.

AonBenfield. 2019. *Weather, Climate & Catastrophe Insight: 2018 Annual Report*. 2019. http://thoughtleadership.aonbenfield.com/Documents/20190122-ab-if-annual-weather-climate-report-2018.pdf.

APS. 2011. *Direct Air Capture of CO_2 with Chemicals*. American Physical Society. 2011. https://www.aps.org/policy/reports/assessments/upload/dac2011.pdf.

Arcadia Power. 2014. "15 Things Americans Spend More on Per Year Than the EPA Budget." 2014. https://blog.arcadiapower.com/15-things-american-spend-per-year-entire-epa-budget/.

Aronoff, Kate. 2019. "How Much Destruction Is Needed for Us to Take Climate Change Seriously?" 2019. https://www.theguardian.com/commentisfree/2019/sep/03/how-much-destruction-is-needed-for-us-to-take-climate-change-seriously.

Arora, V. K., and G. J. Boer. 2014. "Terrestrial Ecosystems Response to Future Changes in Climate and Atmospheric CO_2 Concentration." *Biogeosciences* 11, no. 15 (2014): 4157–71. https://doi.org/10.5194/bg-11-4157-2014.

Arora, V. K., and J. F. Scinocca. 2016. "Constraining the Strength of the Terrestrial CO_2 Fertilization Effect in the Canadian Earth System Model Version 4.2 (CanESM4.2)." *Geoscientific Model Development* 9, no. 7 (2016.): 2357–76. https://doi.org/10.5194/gmd-9-2357-2016.

Arora, Vivek K., and Joe R. Melton. 2018. "Reduction in Global Area Burned and Wildfire Emissions since 1930s Enhances Carbon Uptake by Land." *Nature Communications* 9, no. 1 (2018): 1326. https://doi.org/10.1038/s41467-018-03838-0.

Arrhenius, Svante. 1896. "On the Influence of Carbonic Acid in the Air upon the Temperature of the Ground." *London, Edinburgh, and Dublin Philosophical Magazine and Journal of Science* 41 (1896): 237–75.

Ashley, Walker S., Stephen Strader, Troy Rosencrants, and Andrew J. Krmenec. 2014. "Spatiotemporal Changes in Tornado Hazard Exposure: The Case of the Expanding Bull's-Eye Effect in Chicago, Illinois." *Weather, Climate, and Society* 6, no. 2 (2014): 175–93. https://doi.org/10.1175/WCAS-D-13-00047.1.

Astor, Maggie. 2018. "No Children Because of Climate Change? Some People Are Considering It." 2018. https://www.nytimes.com/2018/02/05/climate/climate-change-children.html.

Athawes, Simon. 2018. "Five Years On: How Haiyan Shocked the World." https://www.rms.com/blog/2018/11/08/five-years-on-how-haiyan-shocked-the-world/.

Auffhammer, M., and E. T. Mansur. 2014. "Measuring Climatic Impacts on Energy Consumption: A Review of the Empirical Literature." *Energy Economics* 46 (2014): 522–30. https://doi.org/10.1016/j.eneco.2014.04.017.

BA. 2006. "Click for the Climate." *BA National Sience Week 2006.* 2006. https://web.archive.org/web/20080906185155if_/http://www.the-ba.net/the-ba/Events/NSEW/AboutNSEW/NSEW_archive/NationalScienceWeek2006/ClimateChange/_ClickfortheClimate.htm.

Baddour, Dylan. 2017. "High-Tech Flood Study Calls for Buyouts of North Houston Apartments." 2017. https://www.houstonchronicle.com/business/article/High-tech-flood-study-calls-for-buyouts-of-North-11198088.php.

Bailey, Ronald. 2013. "Peak Farmland?" 2013. https://reason.com/2013/05/27/peak-farmland/.

Bainimarama, Josaia Voreqe. 2013. "PM's Speech at the Opening of the Rukua Smart Solar Power Project." 2013. http://www.ruraldev.gov.fj/index.php/19-frontpage-articles/75-commissioning-of-tissue-culture-lab-at-koronivia-research-station.

Bakhsh, Khuda, and M. Asif Kamran. 2019. "Adaptation to Climate Change in Rain-Fed Farming System in Punjab, Pakistan." *International Journal of the Commons* 13, no. 2 (2019): 833–47. https://doi.org/10.5334/ijc.887.

Bakkensen, Laura A., and Robert O. Mendelsohn. 2016. "Risk and Adaptation: Evidence from Global Hurricane Damages and Fatalities." *Journal of the Association of Environmental and Resource Economists* 3, no. 3 (2016): 555–87. https://doi.org/10.1086/685908.

Bamat, Joseph. 2015. "France Takes Steps to Avoid Repeat of Deadly 2003 Heat Wave." 2015. https://www.france24.com/en/20150701-france-paris-heat-wave-alert-deadly-2003-summer-guidelines.

Bamber, Jonathan L., Michael Oppenheimer, Robert E. Kopp, Willy P. Aspinall, and Roger M. Cooke. 2019. "Ice Sheet Contributions to Future Sea-Level Rise from Structured Expert Judgment." *Proceedings of the National Academy of Sciences* 116, no. 23 (2019): 11195–200. https://doi.org/10.1073/pnas.1817205116.

Banerjee, Lopamudra. 2010. "Effects of Flood on Agricultural Productivity in Bangladesh." *Oxford Development Studies* 38, no. 3 (2010): 339–56. https://doi.org/10.1080/13600818.2010.505681.

Barcelona Institute for Global Health. 2018. "Decline in Heat-Related Deaths in Spain despite Rising Summer Temperatures." 2018. https://www.eurekalert.org/pub _releases/2018-07/bifg-dih071818.php.

Baron, Jessica. 2019. "Flight-Shaming Is Now a Thing—Will It Keep You from Traveling?" 2019. https://www.forbes.com/sites/jessicabaron/2019/07/02/flight -shaming-is-now-a-thing--will-it-keep-you-from-traveling/#60eaa47b71a3.

Baskin, Jeremy. 2019. *Geoengineering, the Anthropocene and the End of Nature*. London: Palgrave Macmillan. https://search.ebscohost.com/login.aspx?direct=true &scope=site&db=nlebk&db=nlabk&AN=2140070.

Bawden, Tom. 2016. "COP21: Paris Deal Far Too Weak to Prevent Devastating Climate Change, Academics Warn." 2016. https://www.independent.co.uk/environment /climate-change/cop21-paris-deal-far-too-weak-to-prevent-devastating-climate -change-academics-warn-a6803096.html.

BEA. 2019. "Table 1.1.9. Implicit Price Deflators for Gross Domestic Product." Bureau of Economic Analysis. 2019. https://apps.bea.gov/iTable/index_nipa.cfm.

Belfast Telegraph. 2014. "OAPs Staying in Bed to Keep Warm." https://www.belfast telegraph.co.uk/news/uk/oaps-staying-in-bed-to-keep-warm-29890928.html.

Benichou, Léo. 2014. *E-Data of Etemad and Luciani, World Energy Production 1800—1985*. http://www.tsp-data-portal.org/Energy-Production-Statistics#tspQvChart.

Berke, Philip R. 2017. "Why Is Houston So Vulnerable to Devastating Floods?" 2017. https://www.bbc.com/news/world-us-canada-41107049.

Bern, C., J. Sniezek, G. M. Mathbor, M. S. Siddiqi, C. Ronsmans, A. M. Chowdhury, A. E. Choudhury, K. Islam, M. Bennish, and E. Noji. 1993. "Risk Factors for Mortality in the Bangladesh Cyclone of 1991." *Bulletin of the World Health Organization* 71, no. 1 (1993): 73–78.

Berners-Lee, M., C. Hoolohan, H. Cammack, and C. N. Hewitt. 2012. "The Relative Greenhouse Gas Impacts of Realistic Dietary Choices." *Energy Policy* 43 (2012): 184–90. https://doi.org/10.1016/j.enpol.2011.12.054.

Berrigan, Frida. 2019. "Parenting in a World Hurtling Toward Catastrophe." 2019. https://www.thenation.com/article/parenting-generation-hot-global-warming -climate-change/.

Berrittella, M., A. Bigano, R. Roson, and R. S. J. Tol. 2006. "A General Equilibrium Analysis of Climate Change Impacts on Tourism." *Tourism Management* 27, no. 5 (2006): 913–24. https://doi.org/10.1016/j.tourman.2005.05.002.

Berthélemy, Michel, and Lina Escobar Rangel. 2015. "Nuclear Reactors' Construction Costs: The Role of Lead-Time, Standardization and Technological Progress." *Energy Policy* 82 (July 2015): 118–30. https://doi.org/10.1016/j.enpol.2015.03.015.

Bickel, J. Eric, and Lee Lane. 2009. "Fix the Climate: Climate Engineering Assessment." 2009. https://www.copenhagenconsensus.com/publication/fix-climate-climate -engineering-assessment-bickel-lane.

Biello, David. 2014. "Cool Roofs Might Be Enough to Save Cities from Climate Overheating." 2014. https://www.scientificamerican.com/article/cool-roofs-might-be -enough-to-save-cities-from-climate-overheating/.

Biribo, Naomi, and Colin D. Woodroffe. 2013. "Historical Area and Shoreline Change of

Reef Islands around Tarawa Atoll, Kiribati." *Sustainability Science* 8, no. 3 (2013): 345–62. https://doi.org/10.1007/s11625-013-0210-z.

Biswas, Soutik. 2014. "India's Dark History of Sterilisation." *BBC News*, sec. India, November 14, 2014. https://www.bbc.com/news/world-asia-india-30040790.

Bjelle, Eivind Lekve, Kjartan Steen-Olsen, and Richard Wood. 2018. "Climate Change Mitigation Potential of Norwegian Households and the Rebound Effect." *Journal of Cleaner Production* 172 (2018): 208–17. https://doi.org/10.1016/j.jclepro.2017.10.089.

BNEF. 2019. "Energy Storage Investments as Boom Battery Costs Halve in the Next Decade." BloombergNEF. July 31, 2019. https://about.bnef.com/blog/energy-storage-investments-boom-battery-costs-halve-next-decade/.

Bohringer, Christoph, Thomas F. Rutherford, and Richard S. J. Tol. 2009. "The EU 20/20/2020 Targets: An Overview of the EMF22 Assessment." *Energy Economics* International, U.S. and E.U. Climate Change Control Scenarios: Results from EMF 22, 31, Supplement 2 (December 2009): S268–73. https://doi.org/10.1016/j.eneco.2009.10.010.

Bolt, Jutta, Robert Inklaar, Herman de Jong, and Jan Luiten van Zanden. 2018. *"Rebasing 'Maddison'": New Income Comparisons and the Shape of Long-Run Economic Development.* Maddison Project Working Paper 10. 2018. https://www.rug.nl/ggdc/historicaldevelopment/maddison/releases/maddison-project-database-2018.

Bonjour, Sophie, Heather Adair-Rohani, Jennyfer Wolf, Nigel G. Bruce, Sumi Mehta, Annette Prüss-Ustün, Maureen Lahiff, Eva A. Rehfuess, Vinod Mishra, and Kirk R. Smith. 2013. "Solid Fuel Use for Household Cooking: Country and Regional Estimates for 1980–2010." *Environmental Health Perspectives* 121, no. 7 (2013): 784–90. https://doi.org/10.1289/ehp.1205987.

Borenstein, Severin, and Lucas Davis. 2015. "The Distributional Effects of U.S. Clean Energy Tax Credits." w21437. Cambridge, MA: National Bureau of Economic Research. 2015. https://doi.org/10.3386/w21437.

Bos, Frits, and Peter Zwaneveld. 2017. "Cost-Benefit Analysis for Flood Risk Management and Water Governance in the Netherlands: An Overview of One Century." SSRN Scholarly Paper ID 3023983. Rochester, NY: Social Science Research Network, 2017. https://www.cpb.nl/sites/default/files/omnidownload/CPB-Backgrounddocument-August2017-Cost-benefit-analysis-for-flood-risk-management-and-water-governance-overview.pdf.

Bouwer, Laurens M., and Sebastiaan N. Jonkman. 2018. "Global Mortality from Storm Surges Is Decreasing." *Environmental Research Letters* 13, no. 1 (2018). https://doi.org/10.1088/1748-9326/aa98a3.

Bowden, Ebony. 2019. "Climate Change Could End Human Civilization by 2050: Report." 2019. https://nypost.com/2019/06/03/climate-change-could-end-human-civilization-by-2050-report/.

BP. 2019. "BP Statistical Review of World Energy 2019." 2019. https://www.bp.com/content/dam/bp/business-sites/en/global/corporate/pdfs/energy-economics/statistical-review/bp-stats-review-2019-full-report.pdf.

Bradford, Tim. 2017. "Geoengineering Could Create More Problems Than It Could Solve." EcoWatch. November 24, 2017. https://www.ecowatch.com/geoengineering-more-harm-than-good-2512157242.html.

Brady, Dennis, and Chris Mooney. 2019. "More Billion-Dollar US Disasters as World Warms." 2019. https://www.bostonglobe.com/news/nation/2019/02/06/more-billion-dollar-disasters-world-warms/NLkzKhoAd3kqeuAfCbcbbK/story.html.

Branson, Richard. 2019. "Virgin Earth Challenge." Virgin. December 19, 2019. https://www.virgin.com/content/virgin-earth-challenge-0.

Braverman, Blair. 2019. "On Having Kids at the End of the World." 2019. https://www.outsideonline.com/2395047/having-kids-climate-change-ethics.

Broughel, James, and W. Kip Viscusi. 2017. "Death by Regulation: How Regulations Can Increase Mortality Risk." 2017. https://www.mercatus.org/publications/regulations-increase-mortality-risk.

Brulle, Robert J. 2018. "The Climate Lobby: A Sectoral Analysis of Lobbying Spending on Climate Change in the USA, 2000 to 2016." *Climatic Change* 149, no. 3 (2018): 289–303. https://doi.org/10.1007/s10584-018-2241-z.

Bryant, Benjamin P., and Anthony L. Westerling. 2014. "Scenarios for Future Wildfire Risk in California: Links between Changing Demography, Land Use, Climate, and Wildfire." *Environmetrics* 25, no. 6, SI (2014): 454–71. https://doi.org/10.1002/env.2280.

Burgess, Matthew G., Justin Ritchie, John Shapland, and Roger Pielke. 2020. "IPCC Baseline Scenarios Over-Project CO_2 Emissions and Economic Growth." Preprint. SocArXiv. https://doi.org/10.31235/osf.io/ahsxw.

Caldeira, Ken. 2019. "Battery Installation in 2018." 2019. https://twitter.com/KenCaldeira/status/1102986556346757120.

Calvin, Katherine, Leon Clarke, Volker Krey, Geoffrey Blanford, Kejun Jiang, Mikiko Kainuma, Elmar Kriegler, Gunnar Luderer, and P. R. Shukla. 2012. "The Role of Asia in Mitigating Climate Change: Results from the Asia Modeling Exercise." *Energy Economics* 34, Supplement 3 (December 2012): S251–60. https://doi.org/10.1016/j.eneco.2012.09.003.

Calvin, Katherine, Allen Fawcett, and Jiang Kejun. 2012. "Comparing Model Results to National Climate Policy Goals: Results from the Asia Modeling Exercise." *Energy Economics* 34 (December 2012): S306–15. https://doi.org/10.1016/j.eneco.2012.03.008.

Calzadilla, Alvaro, Katrin Rehdanz, Richard Betts, Pete Falloon, Andy Wiltshire, and Richard S. J. Tol. 2013. "Climate Change Impacts on Global Agriculture." *Climatic Change* 120, no. 1–2 (2013): 357–74. https://doi.org/10.1007/s10584-013-0822-4.

Cama, Timothy. 2015. "Obama Joins 19 Countries to Double Clean Energy Funding." 2015. https://thehill.com/policy/energy-environment/261441-obama-joins-18-countries-to-double-clean-energy-funding.

Campagnolo, Lorenza, and Marinella Davide. 2019. "Can the Paris Deal Boost SDGs Achievement? An Assessment of Climate Mitigation Co-Benefits or Side-Effects on Poverty and Inequality." *World Development* 122 (October 2019): 96–109. https://doi.org/10.1016/j.worlddev.2019.05.015.

Canberra Times. 1977. *UN Environment Day: Cancers "Environmental."*1977. http:// nla.gov.au/nla.news-page12231167.

Carbon Independent. 2019. "Emissions from Home Energy Use." 2019. https://www .carbonindependent.org/15.html.

Carrington, Damian. 2017. "Want to Fight Climate Change? Have Fewer Children." *Guardian,* 2017. https://www.theguardian.com/environment/2017/jul/12/want-to-fight-climate-change-have-fewer-children.

———. 2018. "Paul Ehrlich: 'Collapse of Civilisation Is a Near Certainty within Decades.'" *Guardian,* sec. Cities. March 22, 2018. https://www.theguardian.com /cities/2018/mar/22/collapse-civilisation-near-certain-decades-population -bomb-paul-ehrlich.

———. 2019. "Why the Guardian Is Changing the Language It Uses about the Environment." *Guardian,* May 17, 2019.

Carson, Rachel. 1962. *Silent Spring.* Boston: Houghton Mifflin, 1962.

Carter, Colin, Xiaomeng Cui, Dalia Ghanem, and Pierre Mérel. 2018. "Identifying the Economic Impacts of Climate Change on Agriculture." *Annual Review of Resource Economics* 10, no. 1 (2018): 361–80. https://doi.org/10.1146/annurev -resource-100517-022938.

Caspani, Marla. 2019. "U.S. Democrats and Republicans Can't Even Agree on the Weather: Reuters/Ipsos." *Reuters.* 2019. https://www.reuters.com/article/us-usa -climate-poll/u-s-democrats-and-republicans-cant-even-agree-on-the-weather -reuters-ipsos-idUSKCN1UK1FY.

CAT. 2018. "Climate Action Tracker Data." 2018. https://climateactiontracker.org /documents/509/CAT_2018-12-09_PublicData_EmissionsPathways_Dec2018 update.xls.

CAT India. 2016. *Climate Action Tracker 2016 India Assessment.* 2016. https://climate actiontracker.org/media/documents/2018/4/CAT_2016-11-02_CountryAssessment _India.pdf.

CBS News. 2006. "2006: Al Gore Does Sundance." 2006. https://www.cbsnews.com /news/2006-al-gore-does-sundance/.

Census, ed. 1975. *Historical Statistics of the United States: Colonial Times to 1970.* Washington, DC: US Government Printing Office, 1975. https://www.census.gov /library/publications/1975/compendia/hist_stats_colonial-1970.html.

———. 1992. *Population of States and Counties of the United States: 1790 to 1990.* 1992. https://www.census.gov/population/www/censusdata/pop1790-1990 .html.

———. 2010. *Coastline Population Trends in the United States: 1960 to 2008.* 2010. https://www.census.gov/library/publications/2010/demo/p25-1139.html.

———. 2011. *Historical Census of Housing Tables.* 2011. https://www.census.gov/hhes /www/housing/census/historic/units.html.

———. 2012. *USA Counties: 2011, Population.* 2012. https://www.census.gov/library /publications/2011/compendia/usa-counties-2011.html.

———. 2017. *2017 National Population Projections Tables.* 2017. https://www.census .gov/data/tables/2017/demo/popproj/2017-summary-tables.html.

————. 2018a. *Annual Estimates of Housing Units for the United States, Regions, Divisions, States, and Counties: April 1, 2010 to July 1, 2017.* 2018. https://factfinder .census.gov/bkmk/table/1.0/en/PEP/2017/PEPANNHU.

————. 2018b. *Median and Average Sale Price of Houses Sold.* 2018. https://www .census.gov/construction/nrs/historical_data/index.html.

Centre for Research on the Epidemiology of Disasters. 2019. "Emergency Events Database (EM-DAT)." 2019. https://www.emdat.be/.

Chakrabortty, Aditya. 2008. "Secret Report: Biofuel Caused Food Crisis." 2008. https:// www.theguardian.com/environment/2008/jul/03/biofuels.renewableenergy.

Challinor, A. J., J. Watson, D. B. Lobell, S. M. Howden, D. R. Smith, and N. Chhetri. 2014. "A Meta-Analysis of Crop Yield under Climate Change and Adaptation." *Nature Climate Change* 4, no. 4 (2014): 287–91. https://doi.org/10.1038/nclimate2153.

Chambers, Dustin, Courtney A. Collins, and Alan Krause. 2019. "How Do Federal Regulations Affect Consumer Prices? An Analysis of the Regressive Effects of Regulation." *Public Choice* 180, no. 1–2 (2019): 57–90. https://doi.org/10.1007/s11127-017 -0479-z.

Chen, Chi, Taejin Park, Xuhui Wang, Shilong Piao, Baodong Xu, Rajiv K. Chaturvedi, Richard Fuchs, et al. 2019. "China and India Lead in Greening of the World through Land-Use Management." *Nature Sustainability* 2, no. 2 (2019): 122–29. https://doi .org/10.1038/s41893-019-0220-7.

Chen, Minjie, Bruno Wichmann, Marty Luckert, Leigh Winowiecki, Wiebke Förch, and Peter Läderach. 2018. "Diversification and Intensification of Agricultural Adaptation from Global to Local Scales." *PLOS ONE* 13, no. 5 (2018): e0196392. https://doi .org/10.1371/journal.pone.0196392.

Chen, Wenfang, Yi Lu, Shao Sun, Yihong Duan, and Gregor C. Leckebusch. 2018. "Hazard Footprint-Based Normalization of Economic Losses from Tropical Cyclones in China During 1983–2015." *International Journal of Disaster Risk Science* 9, no. 2 (2018): 195–206. https://doi.org/10.1007/s13753-018-0172-y.

China NDC. 2016. "China NDC Submission." 2016. https://www4.unfccc.int/sites /ndcstaging/PublishedDocuments/China%20First/China%27s%20First%20NDC %20Submission.pdf.

Chirakijja, Janjala, Seema Jayachandran, and Pinchuan Ong. 2019. "Inexpensive Heating Reduces Winter Mortality." 2019. https://www.nber.org/papers/w25681.

Chiras, Daniel D. 1998. *Environment Science with Infotrac: A Systems Approach to Sustainable Development.* Wadsworth, 1998.

Christensen, P., K. Gillingham, and W. Nordhaus. 2018. "Uncertainty in Forecasts of Long-Run Economic Growth." *Proceedings of the National Academy of Sciences* 115, no. 21 (2018): 5409. https://doi.org/10.1073/pnas.1713628115.

City of Atlanta. 2015. "Heat Relief—Cooling Centers Now Open." 2015. https://www .atl311.com/wp-content/uploads/2015/06/City-of-Atlanta-Cooling-Centers.pdf.

City of New York. 2019. "NYC °CoolRoofs." 2019. https://web.archive.org/web /20190630022627/https://www1.nyc.gov/html/gbee/html/initiatives/coolroofs .shtml.

Climate Feedback. 2017. "Scientists Explain What New York Magazine Article on 'The Uninhabitable Earth' Gets Wrong." 2017. https://climatefeedback.org/evaluation

/scientists-explain-what-new-york-magazine-article-on-the-uninhabitable
-earth-gets-wrong-david-wallace-wells/.

―――. 2019a. "Claim That Human Civilization Could End in 30 Years Is Speculative, Not Supported with Evidence." 2019. https://climatefeedback.org/evaluation /iflscience-story-on-speculative-report-provides-little-scientific-context -james-felton/.

―――. 2019b. "Prediction by Extinction Rebellion's Roger Hallam That Climate Change Will Kill 6 Billion People by 2100 Is Unsupported." 2019. https://climatefeedback .org/claimreview/prediction-extinction-rebellion-climate-change-will-kill-6 -billion-people-unsupported-roger-hallam-bbc/.

Climate Home News. 2013. "'It's Time to Stop This Madness'—Philippines Plea at UN Climate Talks," 2013. https://www.climatechangenews.com/2013/11/11/its-time -to-stop-this-madness-philippines-plea-at-un-climate-talks/.

Climate Nexus. 2018. "Experts React to Historic IPCC Report on Limiting Warming to 1.5°C." 2018. https://climatenexus.org/international/experts-react-to-historic -ipcc-report-on-limiting-warming-to-1-5c/.

CNN. 2018. "Earth Has 12 Years to Avert Climate Change Catastrophe, Warns UN Report." 2018. https://edition.cnn.com/videos/tv/2018/10/08/news-stream-stout -skea-ipcc-report-climate-change-2030.cnn.

―――. 2019. "Dozens Dead in One of India's Longest Heat Waves." https://edition .cnn.com/videos/world/2019/06/14/india-heatwave-kk-lon-orig.cnn.

Colborn, Theo, Dianne Dumanoski, and John Peterson Myers. 1996. *Our Stolen Future: Are We Threatening Our Fertility, Intelligence, and Survival?: A Scientific Detective Story.* New York: Dutton, 1996.

Collier, Peter. 1970. ". . . Ecological Destruction Is a Condition of American Life . . . An Interview with Ecologist Paul Ehrlich." *Mademoiselle* 189 (April 1970), 188–189, 291–98.

Colt, Stephen G., and Gunnar P. Knapp. 2016. "Economic Effects of an Ocean Acidification Catastrophe." *American Economic Review* 106, no. 5 (2016): 615–19. https:// doi.org/10.1257/aer.p20161105.

Conforti, Piero, ed. 2011. *Looking Ahead in World Food and Agriculture: Perspectives to 2050.* Rome: Food and Agriculture Organization of the United Nations, 2011.

Connor, Steve. 2014. "Climate Change Is Increasing the Risk of Malaria for People Living in Mountainous Regions in the Tropics." 2014. https://www.independent .co.uk/news/science/climate-change-is-increasing-the-risk-of-malaria-for -people-living-in-mountainous-regions-in-the-9174448.html.

Copenhagen Consensus Center. 2019. "Experts." 2019. https://www.copenhagen consensus.com/expert.

Corner, Adam. 2014. "Will Geoengineering Make People Give up Cutting Their Carbon Footprint?" 2014. https://www.theguardian.com/environment/2014/nov/17 /geoengineering-co2-carbon-dioxide-earth-climate.

Costinot, Arnaud, Dave Donaldson, and Cory Smith. 2016. "Evolving Comparative Advantage and the Impact of Climate Change in Agricultural Markets: Evidence from 1.7 Million Fields around the World." *Journal of Political Economy* 124, no. 1 (2016): 205–48. https://doi.org/10.1086/684719.

CPI. 2019. *Consumer Price Index 1800–*. 2019. https://www.minneapolisfed.org/about-us/monetary-policy/inflation-calculator/consumer-price-index-1800-

Crompton, Ryan P., K. John McAneney, Keping Chen, Roger A. Pielke, and Katharine Haynes. 2010. "Influence of Location, Population, and Climate on Building Damage and Fatalities Due to Australian Bushfire: 1925–2009." *Weather, Climate, and Society* 2, no. 4 (2010): 300–10. https://doi.org/10.1175/2010WCAS1063.1.

Cunningham, William P, and Mary Ann Cunningham. 2017. *Environmental Science: A Global Concern.* New York: McGraw-Hill, NY, 2017.

Dahlgreen, Will. 2016. "Chinese People Are Most Likely to Feel the World Is Getting Better." 2016. https://yougov.co.uk/topics/lifestyle/articles-reports/2016/01/05/chinese-people-are-most-optimistic-world.

Dannenberg, Astrid, and Sonja Zitzelsberger. 2019. "Climate Experts' Views on Geoengineering Depend on Their Beliefs about Climate Change Impacts." *Nature Climate Change* 9, no. 10 (2019): 769–75. https://doi.org/10.1038/s41558-019-0564-z.

Dasgupta, Susmita, Mainul Huq, Zahirul Huq Khan, Sohel Masud, Manjur Murshed Zahid Ahmed, Nandan Mukherjee, and Kiran Pandey. 2011. "Climate Proofing Infrastructure in Bangladesh: The Incremental Cost of Limiting Future Flood Damage." *Journal of Environment & Development* 20, no. 2 (2011): 167–90. https://doi.org/10.1177/1070496511408401.

Davies, Simon. 2018. "Rising Sea Levels Could Cost the World $14 Trillion a Year by 2100." 2018. https://ioppublishing.org/news/rising-sea-levels-cost-world-14-trillion-year-2100/.

Davies, Stephen. 2004. "The Great Horse-Manure Crisis of 1894." 2004. https://fee.org/articles/the-great-horse-manure-crisis-of-1894/.

Davis, Lance E., Robert E. Gallman, and Teresa D. Hutchins. 1988. "The Decline of U.S. Whaling: Was the Stock of Whales Running Out?" *Business History Review* 62, no. 4 (1988): 569–95. https://doi.org/10.2307/3115618.

Davis, Lucas W., and James M. Sallee. 2019. "Should Electric Vehicle Drivers Pay a Mileage Tax?" National Bureau of Economic Research. Working Paper 26072. 2019. https://doi.org/10.3386/w26072.

Davis, Robert E., Paul C. Knappenberger, Patrick J Michaels, and Wendy M. Novicoff. 2003. "Changing Heat-Related Mortality in the United States." *Environmental Health Perspectives* 111, no. 14 (2003): 1712–18. https://doi.org/10.1289/ehp.6336.

Davlasheridze, Meri, Kayode O. Atoba, Samuel Brody, Wesley Highfield, William Merrell, Bruce Ebersole, Adam Purdue, and Robert W. Gilmer. "Economic Impacts of Storm Surge and the Cost-Benefit Analysis of a Coastal Spine as the Surge Mitigation Strategy in Houston-Galveston Area in the USA." *Mitigation and Adaptation Strategies for Global Change* 24, no. 3 (2019): 329–54. https://doi.org/10.1007/s11027-018-9814-z.

De Châtel, Francesca. 2014. "The Role of Drought and Climate Change in the Syrian Uprising: Untangling the Triggers of the Revolution." *Middle Eastern Studies* 50, no. 4 (2014): 521–35. https://doi.org/10.1080/00263206.2013.850076.

Deign, Jason. 2019. "Europe's EPR Nuclear Reactor Model May Finally Go Live in 2019." September 1, 2019. https://www.greentechmedia.com/articles/read/epr-nuclear-reactor-model-may-finally-go-live-europe-2019.

Dennis, Brady, Steven Mufson, and Scott Clement. 2019. "Americans Increasingly See Climate Change as a Crisis, Poll Shows." 2019. https://www.washingtonpost.com /climate-environment/americans-increasingly-see-climate-change-as-a-crisis -poll-shows/2019/09/12/74234db0-cd2a-11e9-87fa-8501a456c003_story.html.

Department for Business, Energy & Industrial Strategy. 2019. "Domestic Energy Price Indices." 2019. https://www.gov.uk/government/statistical-data-sets/monthly -domestic-energy-price-stastics.

Department for International Development, and Alok Sharma. 2019. "Healthcare for Everyone Must Prioritise Women's Sexual and Reproductive Health and Rights, Says UK at UN General Assembly." 2019. https://www.gov.uk/government/news /healthcare-for-everyone-must-prioritise-womens-sexual-and-reproductive -health-and-rights-says-uk-at-un-general-assembly.

Department for International Development, Prime Minister's Office, David Cameron, and Justine Greening. 2013. "World Leaders Sign Global Agreement to Help Beat Hunger." 2013. https://www.gov.uk/government/news/world-leaders-sign-global -agreement-to-help-beat-hunger-and-malnutrition.

Di Falco, Salvatore, Marcella Veronesi, and Mahmud Yesuf. 2011. "Does Adaptation to Climate Change Provide Food Security? A Micro-Perspective from Ethiopia." *American Journal of Agricultural Economics* 93, no. 3 (2011): 829–46. https://doi .org/10.1093/ajae/aar006.

Diaz, J., R. Carmona, I. J. Mirón, C. Ortiz, and C. Linares. 2015. "Comparison of the Effects of Extreme Temperatures on Daily Mortality in Madrid (Spain), by Age Group: The Need for a Cold Wave Prevention Plan." *Environmental Research* 143 (2015): 186–91. https://doi.org/10.1016/j.envres.2015.10.018.

Dickinson, Tim. 2019. "How to Survive a Flooded World." 2019. https://www.rolling stone.com/politics/politics-features/craig-fugate-obama-fema-imelda-flooding -887712/.

Dinda, Soumyananda. 2004. "Environmental Kuznets Curve Hypothesis: A Survey." *Ecological Economics* 49, no. 4 (2004): 431–55. https://doi.org/10.1016/j.ecolecon .2004.02.011.

Doerr, Stefan H., and Cristina Santín. 2016. "Global Trends in Wildfire and Its Impacts: Perceptions versus Realities in a Changing World." *Philosophical Transactions of the Royal Society of London. Series B, Biological Sciences* 371, no. 1696 (2016): 20150345. https://doi.org/10.1098/rstb.2015.0345.

Dolin, Eric Jay. 2008. *Leviathan: The History of Whaling in America.* New York: W.W. Norton, 2008.

Doll, Richard, and Richard Peto. 1981. "The Causes of Cancer: Quantitative Estimates of Avoidable Risks of Cancer in the United States Today." *Journal of the National Cancer Institute* 66, no. 6 (1981): 1191–1308.

Dollar, David, Tatjana Kleineberg, and Aart Kraay. 2016. "Growth Still Is Good for the Poor." *European Economic Review* 81 (2016):68–85. https://doi.org/10.1016/j .euroecorev.2015.05.008.

Donat, M. G., L. V. Alexander, H. Yang, I. Durre, R. Vose, R. J. H. Dunn, K. M. Willett, et al. 2013. "Updated Analyses of Temperature and Precipitation Extreme Indices since the Beginning of the Twentieth Century: The HadEX2 Dataset."

Journal of Geophysical Research-Atmospheres 118, no. 5 (2013): 2098–118. https://doi.org/10.1002/jgrd.50150.

Downton, M, J. Z. B. Miller, and R. A. Pielke. 2005. "Reanalysis of U.S. National Weather Service Flood Loss Database." *Natural Hazards Review* 2, no. 4 (2005): 157–66.

Duan, Hong-Xia, Lü Yan-Li, and Li Yan. 2014. "Chinese Public's Willingness to Pay for CO_2 Emissions Reductions: A Case Study from Four Provinces/Cities." *Advances in Climate Change Research* 5, no. 2 (2014): 100–110. https://doi.org/10.3724/SP.J.1248.2014.100.

Dütschke, Elisabeth, Manuel Frondel, Joachim Schleich, and Colin Vance. 2018. "Moral Licensing—Another Source of Rebound?" *Frontiers in Energy Research* 6 (2018): 38. https://doi.org/10.3389/fenrg.2018.00038.

Duvat, Virginie K. E. 2019. "A Global Assessment of Atoll Island Planform Changes over the Past Decades." WIREs Climate Change 10, no. 1 (2019): e557. https://doi.org/10.1002/wcc.557.

EASA, EEA, and EUROCONTROL. 2019. "European Aviation Environmental Report 2019." 2019. https://ec.europa.eu/transport/sites/transport/files/2019-aviation-environmental-report.pdf.

Economist. 2009. "Flight to Value." 2009. https://www.economist.com/business/2009/08/06/flight-to-value.

———. 2017. "Climate Change and Inequality—The Rich Pollute, the Poor Suffer." 2017. https://www.economist.com/finance-and-economics/2017/07/13/climate-change-and-inequality.

———. 2018. "Air-Conditioners Do Great Good, but at a High Environmental Cost." 2018. https://www.economist.com/international/2018/08/25/air-conditioners-do-great-good-but-at-a-high-environmental-cost.

ECOS. 2007. "New Zealand to Be Carbon Neutral by 2020." 2007. http://www.ecosmagazine.com/?act=view_file&file_id=EC136p7b.pdf.

Editors of Ramparts. 1970. *Eco-Catastrophe*. San Francisco: Canfield Press, 1970.

Edwardes-Evans, Henry. 2019. "Finland's Olkiluoto-3 Nuclear Plant Full Generation Pushed Back to March 2021 | S&P Global Platts." December 19, 2019. https://www.spglobal.com/platts/en/market-insights/latest-news/electric-power/121919-finlands-olkiluoto-3-nuclear-plant-full-generation-pushed-back-to-march-2021.

EEA. 2018. "Interview—The Dutch Make Room for the River." 2018. https://www.eea.europa.eu/signals/signals-2018-content-list/articles/interview-2014-the-dutch-make.

———. 2019. "Climate Change Threatens Future of Farming in Europe." 2019. https://www.eea.europa.eu/highlights/climate-change-threatens-future-of.

Egan, Patrick J., and Megan Mullin. 2017. "Climate Change: US Public Opinion." *Annual Review of Political Science* 20, no. 1 (2017): 209–27. https://doi.org/10.1146/annurev-polisci-051215-022857.

Ehrlich, Paul R. 1968. *The Population Bomb*. New York: Ballantine Books, 1968.

Ehrlich, Paul Ralph. 1971. *The Population Bomb*. Cutchogue, NY: Buccaneer Books, 1971.

EIA. 2012. *Energy Perspectives 2011*. 2012. https://www.eia.gov/totalenergy/data/annual/EnergyPerspectives.xls.

———. 2017a. *Annual Energy Outlook 2017, with Projections to 2050.* 2017. https://www.eia.gov/outlooks/aeo/pdf/0383(2017).pdf.

———. 2017b. "Annual Energy Outlook 2018. Table: Energy-Related Carbon Dioxide Emissions by Sector and Source." 2017. https://www.eia.gov/outlooks/aeo/data/browser/#/?id=17-AEO2018&region=1-0&cases=ref2018&start=2016&end=2050&f=A&linechart=ref2018-d121317a.3-17-AEO2018.1-0~&map=ref2018-d121317a.4-17-AEO2018.1-0&ctype=linechart&sourcekey=0.

———. 2018. *U.S. Energy-Related CO$_2$ Emissions Expected to Rise Slightly in 2018, Remain Flat in 2019.* 2018. https://www.eia.gov/todayinenergy/detail.php?id=34872.

———. 2019a. *Annual Energy Outlook 2019.* 2019. https://www.eia.gov/outlooks/aeo/.

———. 2019b. "Electric Power Monthly." 2019. https://www.eia.gov/electricity/monthly/epm_table_grapher.php?t=epmt_5_06_a.

———. 2019c. *October 2019 Monthly Energy Review.* 2019. https://www.eia.gov/totalenergy/data/browser/index.php?tbl=T01.03#/?f=A.

EIRP. 2017. *What Will Advanced Nuclear Power Plants Cost?* 2017. http://www.innovationreform.org/wp-content/uploads/2018/01/Advanced-Nuclear-Reactors-Cost-Study.pdf.

Eisenhower, Dwight D. 1961. Military-Industrial Complex Speech. 1961. https://avalon.law.yale.edu/20th_century/eisenhower001.asp.

Elliott, Larry, and Ashley Seager. 2007. "Cut Carbon by Up to Third to Save Poor, UN Tells West." 2007. https://www.theguardian.com/environment/2007/nov/28/climatechange.

EM-DAT. 2020. *The International Disaster Database.* 2020. Centre for Research on the Epidemiology of Disasters. https://www.emdat.be/database.

Encyclopedia Britannica, "Norman Ernest Borlaug," 2020. https://www.britannica.com/biography/Norman-Borlaug.

Environmental Action. 1970. *Earth Day: The Beginning: A Guide for Survival.* New York: Arno Press and *New York Times,* 1970.

EPA. 1997. *The Benefits and Costs of the Clean Air Act, 1970 to 1990—Retrospective Study.* 1997. https://www.epa.gov/clean-air-act-overview/benefits-and-costs-clean-air-act-1970-1990-retrospective-study.

EPI. 2018. *2018 Environmental Performance Index.* 2018. https://epi.envirocenter.yale.edu/2018-epi-report/introduction.

EPRS. 2019. *Mainstreaming of Climate Action in the EU Budget, European Parliamentary Research Service.* 2019. http://www.europarl.europa.eu/RegData/etudes/IDAN/2019/642239/EPRS_IDA(2019)642239_EN.pdf.

Estrada, Francisco, W. J. Wouter Botzen, and Richard S. J. Tol. 2017. "A Global Economic Assessment of City Policies to Reduce Climate Change Impacts." *Nature Climate Change* 7, no. 6 (2017): 403–6. https://doi.org/10.1038/nclimate3301.

Etemad, Bouda, and Jean Luciani. 1991. *World Energy Production, 1800–1985 = Production Mondiale d'énergie, 1800–1985.* Publications Du Centre d'Histoire Économique Internationale de l'Université de Genève 7. Genève: Librairie Droz, 1991.

EU. 2019. *Handbook on the External Costs of Transport: Version 2019.* 2019. http://publications.europa.eu/publication/manifestation_identifier/PUB_MI0518051ENN.

EUNDC. 2016. "European Union NDC Submission." 2016. https://www4.unfccc.int
/sites/ndcstaging/PublishedDocuments/European%20Union%20First/LV-03
-06-EU%20INDC.pdf.

European Union. 2003. "Directive 2003/30/EC of the European Parliament and of the
Council of 8 May 2003 on the Promotion of the Use of Biofuels and Other Renewable Fuels for Transport." 2003. https://eur-lex.europa.eu/LexUriServ/LexUriServ
.do?uri=OJ:L:2003:123:0042:0046:EN:PDF.

Eurostat. 2019a. "Electricity Price Statistics." 2019. https://ec.europa.eu/eurostat
/statistics-explained/index.php/Electricity_price_statistics.

———. 2019b. *EU Statistical Database for Public Finance.* 2019. https://ec.europa.eu
/eurostat/web/government-finance-statistics/data/database.

Ewers, Robert M. 2006. "Interaction Effects between Economic Development and
Forest Cover Determine Deforestation Rates." *Global Environmental Change* 16,
no. 2 (2006): 161–69. https://doi.org/10.1016/j.gloenvcha.2005.12.001.

Extinction Rebellion. 2019. "The Truth." https://rebellion.earth/the-truth/. 2019.

Fagerlund, Karin. 2018. "Förra FN-Chefen Vill Förbjuda Kött På Restauranger: 'Vill
de Äta Kött Får de Göra Det Utanför.'" 2018. https://www.svt.se/nyheter/utrikes
/vill-forbjuda-kott-pa-restauranger-kottatare-ska-behandlas-som-rokare-pa
-krogen.

Fankhauser, Sam. 2017. "Adaptation to Climate Change." *Annual Review of Resource
Economics* 9, no. 1 (2017): 209–30. https://doi.org/10.1146/annurev-resource-100
516-033554.

FAO. 2019. "FAOSTAT Database." 2019. http://www.fao.org/faostat/en/#data.

Fawcett, Allen A., Leon C. Clarke, Sebastian Rausch, and John P. Weyant. 2014. "Overview of EMF 24 Policy Scenarios." *Energy Journal* 35, no. 01 (2014). https://doi.org
/10.5547/01956574.35.SI1.3.

Federico, Giovanni. 2005. *Feeding the World: An Economic History of Agriculture,
1800–2000.* Princeton Economic History of the Western World. Princeton: Princeton University Press, 2005.

Ferdous, Ruknul, Anna Wesselink, Luigia Brandimarte, Kymo Slager, Margreet
Zwarteveen, and Giuliano Di Baldassarre. 2019. "The Costs of Living with Floods
in the Jamuna Floodplain in Bangladesh." *Water* 11, no. 6 (2019): 1238. https://doi
.org/10.3390/w11061238.

Ferguson, Alex P., and Walker S. Ashley. 2017. "Spatiotemporal Analysis of Residential
Flood Exposure in the Atlanta, Georgia Metropolitan Area." *Natural Hazards* 87,
no. 2 (2017): 989–1016. https://doi.org/10.1007/s11069-017-2806-6.

Fishbach, Ayelet, and Ravi Dhar. 2005. "Goals as Excuses or Guides: The Liberating
Effect of Perceived Goal Progress on Choice." *Journal of Consumer Research* 32,
no. 3 (2005): 370–77. https://doi.org/10.1086/497548.

Flavelle, Christopher. 2018. "California's Wildfire Epidemic Is Blamed on Bad Building
Decisions." 2018. https://www.bloomberg.com/news/articles/2018-11-14/califor
nia-s-wildfire-epidemic-blamed-on-bad-building-decisions.

———. 2019. "Even as Floods Worsen with Climate Change, Fewer People Insure
Against Disaster." 2019. https://www.nytimes.com/2019/06/08/climate/climate
-flood-insurance.html.

Ford, Murray R., and Paul S. Kench. 2015. "Multi-Decadal Shoreline Changes in Response to Sea Level Rise in the Marshall Islands." *Anthropocene* 11 (September 2015): 14–24. https://doi.org/10.1016/j.ancene.2015.11.002.

Ford, Peter. 2019. "Heat Waves: How France Has Cut Death Toll 90% since 2003." *Christian Science Monitor,* November 4, 2019. https://www.csmonitor.com/World /Europe/2019/1104/Heat-waves-How-France-has-cut-death-toll-90-since-2003.

Formetta, Giuseppe, and Luc Feyen. 2019. "Empirical Evidence of Declining Global Vulnerability to Climate-Related Hazards." *Global Environmental Change* 57 (2019): 101920. https://doi.org/10.1016/j.gloenvcha.2019.05.004.

Fouquet, Roger. 2009. "A Brief History of Energy." In *International Handbook on the Economics of Energy*, edited by Lester C. Hunt and Joanne Evans, 1–19. Cheltenham, UK: Edward Elgar, 2009.

Freeman, Ashley C., and Walker S. Ashley. 2017. "Changes in the US Hurricane Disaster Landscape: The Relationship between Risk and Exposure." *Natural Hazards* 88, no. 2 (2017): 659–82. https://doi.org/10.1007/s11069-017-2885-4.

Fricko, Oliver, Petr Havlik, Joeri Rogelj, Zbigniew Klimont, Mykola Gusti, Nils Johnson, Peter Kolp, et al. 2017. "The Marker Quantification of the Shared Socioeconomic Pathway 2: A Middle-of-the-Road Scenario for the 21st Century." *Global Environmental Change* 42 (January 2017): 251–67. https://doi.org/10.1016/j.glo envcha.2016.06.004.

Friedman, Thomas L. 2005. *The World Is Flat: A Brief History of the Twenty-First Century.* First edition. New York: Farrar, Straus and Giroux, 2005.

Fu, Sze Hang, Antonio Gasparrini, Peter S. Rodriguez, and Prabhat Jha. 2018. "Mortality Attributable to Hot and Cold Ambient Temperatures in India: A Nationally Representative Case-Crossover Study." *PLOS Medicine* 15, no. 7 (2018): e1002619. https://doi.org/10.1371/journal.pmed.1002619.

Furness, Hannah. 2019. "We Have 18 Months to Save World, Prince Charles Warns Commonwealth Leaders." 2019. https://www.telegraph.co.uk/royal-family /2019/07/11/18-months-save-world-prince-charles-urges-commonwealth -leaders/.

Furukawa, Chishio. 2014. "Do Solar Lamps Help Children Study? Contrary Evidence from a Pilot Study in Uganda." *Journal of Development Studies* 50, no. 2 (2014): 319–41. https://doi.org/10.1080/00220388.2013.833320.

Galbraith, Kate. 2009. "Having Children Brings High Carbon Impact." 2009. https:// green.blogs.nytimes.com/2009/08/07/having-children-brings-high-carbon -impact/.

Galiana, Isabel, and Amy Sopinka. 2015. "Energy Assessment Paper. Benefits and Cost of the Energy Targets for the Post-2015 Development Agenda." *Copenhagen Consensus Center.* 2015. https://www.copenhagenconsensus.com/sites/default/files /energy_assessment_-_galiana_and_sopinka_0.pdf.

Gallup. 2019. "Crime." 2019. https://news.gallup.com/poll/1603/crime.aspx.

Galvin, J. F. P. 2014. "The Development, Track and Destruction of Typhoon Haiyan." *Weather* 69, no. 11 (2014): 307–9. https://doi.org/10.1002/wea.2458.

Gasparrini, Antonio, Yuming Guo, Masahiro Hashizume, Eric Lavigne, Antonella Zanobetti, Joel Schwartz, Aurelio Tobias, et al. 2015. "Mortality Risk Attributable to

High and Low Ambient Temperature: A Multicountry Observational Study." *Lancet* 386, no. 9991 (2015): 369–75. https://doi.org/10.1016/S0140-6736(14)62114-0.

Gasparrini, Antonio, Yuming Guo, Francesco Sera, Ana Maria Vicedo-Cabrera, Veronika Huber, Shilu Tong, Micheline de Sousa Zanotti Stagliorio Coelho, et al. 2017. "Projections of Temperature-Related Excess Mortality under Climate Change Scenarios." *Lancet Planetary Health* 1, no. 9 (2017): e360–67. https://doi.org/10.1016/S2542-5196(17)30156-0.

Gates, Bill. 2019. "Bill Gates: The Best Investment I've Ever Made." 2019. https://www.wsj.com/articles/bill-gates-the-best-investment-ive-ever-made-11547683309.

Geoengineering Monitor. 2020. "Reasons to Oppose Geoengineering." Geoengineering Monitor. 2020. http://www.geoengineeringmonitor.org/reasons-to-oppose/.

Georgescu, M., P. E. Morefield, B. G. Bierwagen, and C. P. Weaver. 2014. "Urban Adaptation Can Roll Back Warming of Emerging Megapolitan Regions." *Proceedings of the National Academy of Sciences* 111, no. 8 (2014): 2909–14. https://doi.org/10.1073/pnas.1322280111.

Gettelman, A., D. N. Bresch, C. C. Chen, J. E. Truesdale, and J. T. Bacmeister. 2018. "Projections of Future Tropical Cyclone Damage with a High-Resolution Global Climate Model." *Climatic Change* 146, no. 3–4 (2018): 575–85. https://doi.org/10.1007/s10584-017-1902-7.

Gezie, Melese. 2019. "Farmer's Response to Climate Change and Variability in Ethiopia: A Review." Edited by Manuel Tejada Moral. *Cogent Food & Agriculture* 5, no. 1 (2019): 1613770. https://doi.org/10.1080/23311932.2019.1613770.

GFDL/NASA. 2019. *Global Warming and Hurricanes: An Overview of Current Research Results.* https://www.gfdl.noaa.gov/global-warming-and-hurricanes/.

Ghosh, Amitav. 2017. *The Great Derangement: Climate Change and the Unthinkable.* Paperback edition. Randy L. and Melvin R. Berlin Family Lectures. Chicago: University of Chicago Press, 2017.

Global Carbon Project. 2019a. *Global Carbon Budget 2019.* 2019. https://www.globalcarbonproject.org/carbonbudget/index.htm.

———. 2019b. *Terrestrial Emissions for All Nations 1960–2018.* 2019. http://www.globalcarbonatlas.org/en/CO2-emissions.

Global Commission on Adaptation. 2019. "Adapt Now: A Global Call for Leadership on Climate Resilience." 2019. https://gca.org/global-commission-on-adaptation/report.

Goedecke, Theda, Alexander J. Stein, and Matin Qaim. 2018. "The Global Burden of Chronic and Hidden Hunger: Trends and Determinants." *Global Food Security-Agriculture Policy Economics and Environment* 17 (June 2018): 21–29. https://doi.org/10.1016/j.gfs.2018.03.004.

Golden, John, and Hannah Wiseman. 2015. "The Fracking Revolution: Shale Gas as a Case Study in Innovation Policy" 64 (2015): 955–1040.

Goldman, Henry, and Christopher Flavelle. 2019. "Bill de Blasio Seeks to Flood-Proof Lower Manhattan by Adding Land." 2019. https://www.bloomberg.com/news/articles/2019-03-14/de-blasio-seeks-to-flood-proof-lower-manhattan-by-adding-land.

Gomes, Leonard. 2010. *German Reparations, 1919–1932: A Historical Survey*. New York: Palgrave Macmillan, 2010.

Goodell, Jeff. 2019. "The Coming Flood: A Data Error Is Corrected, and Our Future Is Rewritten." 2019. https://www.rollingstone.com/politics/politics-news/sea-level -rise-climate-central-study-906178/.

Gore, Al. 2015. "Statement by Former Vice President Al Gore on the Paris Agreement Reached at the United Nations Framework Convention on Climate Change's 21st Conference of the Parties (COP21)." 2015. https://algore.com/news/statement-by -former-vice-president-al-gore-on-the-paris-agreement-reached-at-the-united -nations-framework-convention-on-climate-change-s-21st-conference-of-the -parties-cop21.

Grabs, Janina. 2015. "The Rebound Effects of Switching to Vegetarianism. A Microeconomic Analysis of Swedish Consumption Behavior." *Ecological Economics* 116 (August 2015): 270–79. https://doi.org/10.1016/j.ecolecon.2015.04.030.

Graham, John D. 1995. "Comparing Opportunities to Reduce Health Risks: Toxin Control, Medicine and Injury Prevention." Policy Report 192. 1995. http://www .ncpathinktank.org/pdfs/st192.pdf.

Gramlich, John. 2019. "5 Facts about Crime in the U.S." 2019. https://www.pew research.org/fact-tank/2019/10/17/facts-about-crime-in-the-u-s/.

Greater London Authority. 2006. *London's Urban Heat Island: A Summary for Decision Makers*. 2006. https://www.puc.state.pa.us/electric/pdf/dsr/dsrwg_sub _ECA-London.pdf.

Greenpeace. 2014. "Tale of Two Worlds." 2014. http://dharnailive.in/stories/tale_of _two_worlds.

Grimm, Michael, Luciane Lenz, Jörg Peters, and Maximiliane Sievert. 2019. "Demand for Off-Grid Solar Electricity—Experimental Evidence from Rwanda." *SSRN Electronic Journal*. 2019. https://doi.org/10.2139/ssrn.3399004.

Grübler, A., B. O'Neill, K. Riahi, V. Chirkov, A. Goujon, P. Kolp, I. Prommer, S. Scherbov, and E. Slentoe. 2007. "Regional, National, and Spatially Explicit Scenarios of Demographic and Economic Change Based on SRES." *Technological Forecasting and Social Change* 74, no. 7 (2007): 980–1029. https://doi.org/10.1016/j.tech fore.2006.05.023.

Grubler, Arnulf. 2010. "The Costs of the French Nuclear Scale-up: A Case of Negative Learning by Doing." *Special Section on Carbon Emissions and Carbon Management in Cities with Regular Papers* 38, no. 9 (2010): 5174–88. https://doi.org /10.1016/j.enpol.2010.05.003.

Grunwald, Michael. 2019. "Climate Change Could Be a Problem in 2020 . . . for Democrats." 2019. https://www.politico.com/magazine/story/2019/09/03/climate -change-democratic-candidates-2020-227910.

Guardian. 2006. "Is Cruising Any Greener than Flying?" 2006. https://www.the guardian.com/travel/2006/dec/20/cruises.green.

———. 2016. "The Guardian and Observer Style Guide." 2016. https://www.the guardian.com/guardian-observer-style-guide-c.

———. 2018. "The EU Needs a Stability and Wellbeing Pact, Not More Growth." 2018.

https://www.theguardian.com/politics/2018/sep/16/the-eu-needs-a-stability -and-wellbeing-pact-not-more-growth.

Guinness. 2019. "First Person Killed by a Car 1896, Guinness World Records." 2019. https://www.guinnessworldrecords.com/world-records/first-person-killed-by -a-car.

Gunatilake, Herath, David Roland-Holst, and Bjorn Larsen. 2016. "Smart Energy Options for Bangladesh." 2016. https://www.copenhagenconsensus.com/sites /default/files/gunitlake_holst_larsen_energy.pdf.

Guo, Mingxin, Weiping Song, and Jeremy Buhain. 2015. "Bioenergy and Biofuels: History, Status, and Perspective." *Renewable and Sustainable Energy Reviews* 42 (February 2015): 712–25. https://doi.org/10.1016/j.rser.2014.10.013.

Gür, Turgut M. 2018. "Review of Electrical Energy Storage Technologies, Materials and Systems: Challenges and Prospects for Large-Scale Grid Storage." *Energy & Environmental Science* 11, no. 10 (2018): 2696–2767. https://doi.org/10.1039 /C8EE01419A.

Gurdus, Lizzy. 2017. "Boeing CEO: Over 80% of the World Has Never Taken a Flight. We're Leveraging That for Growth." 2017. https://www.cnbc.com/2017/12/07 /boeing-ceo-80-percent-of-people-never-flown-for-us-that-means-growth .html.

Gustin, Georgina. 2018. "Climate Change Could Lead to Major Crop Failures in World's Biggest Corn Regions." 2018. https://insideclimatenews.org/news/11062018 /climate-change-research-food-security-agriculture-impacts-corn-vegetables -crop-prices.

Guterres, Antonio. 2018. "At Least $26 Trillion in Economic Benefits and 65 Million New Low-Carbon Jobs Could Be Generated by Ambitious #ClimateAction." 2018. https://twitter.com/antonioguterres/status/1037469082952585216.

Habermeier, H.-U. 2007. "Education and Economy—An Analysis of Statistical Data." *Journal of Materials Education* 29 no. 1–2 (2007): 55–70. https://arxiv.org/pdf /0708.2071.pdf.

Hahn, Robert William, Randall W. Lutter, and W. Kip Viscusi. 2000. *Do Federal Regulations Reduce Mortality?* American Enterprise Institute. 2000.

Hall, Charles A. S. 2009. "The Ehrlichs Strike Again." *BioScience* 59, no. 6 (2009): 522–24. https://doi.org/10.1525/bio.2009.59.6.11.

Hallegatte, Stephane, Colin Green, Robert J. Nicholls, and Jan Corfee-Morlot. 2013. "Future Flood Losses in Major Coastal Cities." *Nature Climate Change* 3 (August 2013): 802.

Hallström, E., A. Carlsson-Kanyama, and P. Börjesson. 2015. "Environmental Impact of Dietary Change: A Systematic Review." *Journal of Cleaner Production* 91 (March 2015): 1–11. https://doi.org/10.1016/j.jclepro.2014.12.008.

Hamilton, Clive. 2015. "Geoengineering Is Not a Solution to Climate Change." 2015. https://www.scientificamerican.com/article/geoengineering-is-not-a-solution -to-climate-change/.

Hance, Jeremy. 2017. "Al Gore and Bangladesh PM Spar over Coal Plants in the Sundarbans." 2017. https://news.mongabay.com/2017/03/al-gore-and-bangladesh-pm -spar-over-coal-plants-in-the-sundarbans/.

Hansen, James E. 2011a. "Baby Lauren and the Kool-Aid." 2011. http://www.columbia
    .edu/%7Ejeh1/mailings/2011/20110729_BabyLauren.pdf.
———. 2011b. *Storms of My Grandchildren: The Truth about the Coming Climate Ca-
    tastrophe and Our Last Chance to Save Humanity*. New York: Bloomsbury, 2011.
———. 2018. "Thirty Years Later, What Needs to Change in Our Approach to Climate
    Change." 2018. https://www.bostonglobe.com/opinion/2018/06/26/thirty-years
    -later-what-needs-change-our-approach-climate-change/dUhizA5ubUSzJLJ
    VZqv6GP/story.html.
Hao, Zengchao, Amir AghaKouchak, Navid Nakhjiri, and Alireza Farahmand. 2014.
    "Global Integrated Drought Monitoring and Prediction System." *Scientific Data*
    1 (March 2014): 140001.
Haque, Ubydul, Masahiro Hashizume, Korine N. Kolivras, Hans J. Overgaard, Bivash
    Das, and Taro Yamamoto. 2012. "Reduced Death Rates from Cyclones in Bangla-
    desh: What More Needs to Be Done?" *Bulletin of the World Health Organization*
    90, no. 2 (2012): 150–56. https://doi.org/10.2471/BLT.11.088302.
Hasegawa, Tomoko, Shinichiro Fujimori, Petr Havlík, Hugo Valin, Benjamin Leon
    Bodirsky, Jonathan C. Doelman, Thomas Fellmann, et al. 2018. "Risk of Increased
    Food Insecurity under Stringent Global Climate Change Mitigation Policy." *Na-
    ture Climate Change* 8, no. 8 (2018): 699–703. https://doi.org/10.1038/s41558-018
    -0230-x.
Hasell, Joe, and Max Roser. 2017. *Famines*. 2017. https://ourworldindata.org/famines.
He, Xiaogang, Yoshihide Wada, Niko Wanders, and Justin Sheffield. 2017. "Intensifi-
    cation of Hydrological Drought in California by Human Water Management."
    Geophysical Research Letters 44, no.4 (2017): 1777–85. https://doi.org/10.1002
    /2016GL071665.
Healey, Jenna. 2016. "Rejecting Reproduction: The National Organization for Non-
    Parents and Childfree Activism in 1970s America." *Journal of Women's History* 28,
    no. 1 (2016): 131–56. https://doi.org/10.1353/jowh.2016.0008.
Henriksen, Pia. 2019. "Klimaangst Har Bredt Sig Til Børn Helt Ned i 1. Klasse." 2019.
    https://skoleliv.dk/debat/art7240676/Klimaangst-har-bredt-sig-til-b%C3%B8rn
    -helt-ned-i-1.-klasse.
Herper, Matthew. 2017. "Can Algae Replace Oil Wells? Craig Venter and Exxon Take a
    Step toward Saying 'Yes.'" 2017. https://www.forbes.com/sites/matthewherper
    /2017/06/19/can-algae-replace-oil-wells-craig-venter-and-exxon-take-a-step
    -toward-saying-yes/#6ca049221226.
Heutel, Garth, Nolan Miller, and David Molitor. 2017. "Adaptation and the Mortality
    Effects of Temperature Across U.S. Climate Regions." w23271. Cambridge, MA: Na-
    tional Bureau of Economic Research, 2017. https://doi.org/10.3386/w23271.
Hickman, Leo. 2010. "James Lovelock: Humans Are Too Stupid to Prevent Climate
    Change." *Guardian*, sec. Environment, March 29, 2010. https://www.theguardian
    .com/science/2010/mar/29/james-lovelock-climate-change.
Hicks, Robert L., ed. 2008. *Greening Aid? Understanding the Environmental Impact of
    Development Assistance*. Oxford: Oxford University Press, 2008.
Hills, Jeremy M., Evanthie Michalena, and Konstantinos J. Chalvatzis. 2018. "Innova-
    tive Technology in the Pacific: Building Resilience for Vulnerable Communities."

Technological Forecasting and Social Change 129 (April 2018): 16–26. https://doi
.org/10.1016/j.techfore.2018.01.008.

Hinkel, Jochen, Daniel Lincke, Athanasios T. Vafeidis, Mahé Perrette, Robert James
Nicholls, Richard S. J. Tol, Ben Marzeion, Xavier Fettweis, Cezar Ionescu, and An-
ders Levermann. 2014. "Coastal Flood Damage and Adaptation Costs under 21st
Century Sea-Level Rise." *Proceedings of the National Academy of Sciences* 111, no. 9
(2014): 3292–97. https://doi.org/10.1073/pnas.1222469111.

Hintermann, Beat. 2016. "Pass-Through of CO_2 Emission Costs to Hourly Electricity
Prices in Germany." *Journal of the Association of Environmental and Resource
Economists* 3, no. 4 (2016): 857–91. https://doi.org/10.1086/688486.

Hodgetts, Katie. 2019. "Boulton, Boris and Brendan, You Are Responsible for This
Climate Crisis." 2019. https://www.joe.co.uk/comment/adam-boulton-boris
-johnson-brendan-oneill-climate-change-229022.

Hollingsworth, Julia. 2019. "Climate Change Could Pose 'Existential Threat' by 2050:
Report." 2019. https://edition.cnn.com/2019/06/04/health/climate-change-exis
tential-threat-report-intl/index.html.

Holthaus, Eric. 2018. "U.N. Climate Report Shows Civilization Is at Stake If We Don't
Act Now." 2018. https://grist.org/article/scientists-calmly-explain-that-civiliza
tion-is-at-stake-if-we-dont-act-now/.

Horton, Susan, and John Hoddinott. 2018. "Benefits and Costs of the Food and Nutrition Targets for the Post-2015 Development Agenda." In *Prioritizing Development*, edited by Bjorn Lomborg, 367–74. Cambridge, UK: Cambridge University
Press, 2018. https://www.copenhagenconsensus.com/publication/post-2015
-consensus-food-security-and-nutrition-perspective-horton-hoddinott.

Hossain, Naomi. 2018. "The 1970 Bhola Cyclone, Nationalist Politics, and the Subsistence Crisis Contract in Bangladesh." *Disasters* 42, no. 1 (2018): 187–203. https://
doi.org/10.1111/disa.12235.

Houston Chronicle. 2017. "Another Flood," 2017. https://www.houstonchronicle.com
/opinion/editorials/article/Another-flood-10867145.php.

Hulme, Mike. 2018. "Against Climate Emergency." 2018. https://mikehulme.org
/against-climate-emergency/.

Humane Research Council. 2014. "Study of Current and Former Vegetarians and Vegans." 2014. https://faunalytics.org/wp-content/uploads/2015/06/Faunalytics
_Current-Former-Vegetarians_Full-Report.pdf.

Hurtt, G. C., L. P. Chini, S. Frolking, R. A. Betts, J. Feddema, G. Fischer, J. P. Fisk, et al.
2011. "Harmonization of Land-Use Scenarios for the Period 1500–2100: 600 Years
of Global Gridded Annual Land-Use Transitions, Wood Harvest, and Resulting
Secondary Lands." *Climatic Change* 109, no. 1 (2011): 117. https://doi.org/10.1007
/s10584-011-0153-2.

IATA. 2018. "Fact Sheet Climate Change & CORSIA." 2018. https://www.iata.org/press
room/facts_figures/fact_sheets/Documents/fact-sheet-climate-change.pdf.

ICAP. 2019. "Allowance Price Explorer." 2019. https://icapcarbonaction.com/en/ets
-prices.

IEA. 2014a. "Africa Energy Outlook." 2014. https://web.archive.org/web/201410210202

04/http://www.iea.org/publications/freepublications/publication/WEO2014_
AfricaEnergyOutlook.pdf.

———. 2014b. "Electric Power Consumption (KWh per Capita)—China." 2014. https://
data.worldbank.org/indicator/EG.USE.ELEC.KH.PC?locations=CN.

———. 2015. "Global EV Outlook 2015." 2015. https://web.archive.org/web/2017
1202223535/https://www.iea.org/media/topics/transport/GlobalEV_Outlook
2015Update_1page.pdf.

———. 2017. "Energy Access Outlook 2017. From Poverty to Prosperity." 2017. https://
www.iea.org/publications/freepublications/publication/WEO2017Special
Report_EnergyAccessOutlook.pdf.

———. 2018. "World Energy Outlook 2018." 2018. https://www.iea.org/weo2018/.

———. 2019a. *Energy Prices and Taxes for OECD Countries*. 2019.

———. 2019b. *Global EV Outlook 2019*. Paris, France. 2019. www.iea.org/publications
/reports/globalevoutlook2019.

———. 2019c. *IEA Energy Technology RD&D Statistics*. 2019. https://doi.org/10.1787
/data-00488-en.

———. 2019d. *IEA World Energy Statistics and Balances*. 2019. https://doi.org/10.1787
/enestats-data-en.

———. 2019e. *Renewables 2019—Analysis and Forecast to 2024*. 2019.

———. 2019f. *World Energy Investment 2019*. 2019. https://webstore.iea.org/world
-energy-investment-2019.

———. 2019g. *World Energy Outlook 2019*. 2019. https://www.iea.org/weo2019/.

———. 2020. *IEA Subsidies Database 2000—2018*. 2020. https://www.iea.org/topics
/energy-subsidies.

IGS. 2019. "How Much Electricity Do My Home Appliances Use?" 2019. https://www
.igs.com/energy-resource-center/Energy-101/how-much-electricity-do-my
-home-appliances-use.

IIASA. 2018. *SSP Database Version 2.0*. 2018. https://tntcat.iiasa.ac.at/SspDb.

Institute for Health Metrics and Evaluation. 2017. "Household Air Pollution from Solid
Fuels." 2017. http://ihmeuw.org/4vvu.

———. 2019. "GBD 2017." 2019. https://vizhub.healthdata.org/gbd-compare/#settings
=effbd3025564a7d86e0b5226e963c5351c445453.

Institutet för nervächstudier. 2018. "6th International Degrowth Conference." 2018.
https://degrowth.se/key-information.

IPCC. 2010. "Statement on IPCC Principles and Procedures." 2010. https://www.ipcc
.ch/site/assets/uploads/2018/04/ipcc-statement-principles-procedures-02-2010
.pdf.

———. 2013a. *Climate Change 2013: The Physical Science Basis. Contribution of Working Group I to the Fifth Assessment Report of the Intergovernmental Panel on Climate Change*. Cambridge, UK: Cambridge University Press, 2013. www.climate
change2013.org.

———. 2013b. "Summary for Policymakers." In *Climate Change 2013: The Physical
Science Basis. Contribution of Working Group I to the Fifth Assessment Report of
the Intergovernmental Panel on Climate Change*, edited by T. F. Stocker, D. Qin,

G.-K. Plattner, M. Tignor, S. K. Allen, J. Boschung, A. Nauels, Y. Xia, V. Bex, and P. M. Midgley, 1–30. Cambridge, UK: Cambridge University Press, 2013. www.climatechange2013.org.

———. 2014a. *Climate Change 2014: Impacts, Adaptation, and Vulnerability*. Cambridge, UK: Cambridge University Press. 2014. https://www.ipcc.ch/report/ar5/wg2/.

———. ed. 2014b. *Climate Change 2014: Mitigation of Climate Change: Working Group III Contribution to the Fifth Assessment Report of the Intergovernmental Panel on Climate Change*. New York: Cambridge University Press, 2014.

———. 2014c. *Key Economic Sectors and Services. In: Climate Change 2014: Impacts, Adaptation, and Vulnerability. Part A: Global and Sectoral Aspects. Contribution of Working Group II to the Fifth Assessment Report of the Intergovernmental Panel on Climate Change*. 2014. https://www.ipcc.ch/site/assets/uploads/2018/02/WGIIAR5-Chap10_FINAL.pdf.

———. 2018. *Global Warming of 1.5°C*. 2018. http://www.ipcc.ch/report/sr15/.

Irvine, Peter, Kerry Emanuel, Jie He, Larry W. Horowitz, Gabriel Vecchi, and David Keith. 2019. "Halving Warming with Idealized Solar Geoengineering Moderates Key Climate Hazards." *Nature Climate Change* 9, no. 4 (2019): 295–99. https://doi.org/10.1038/s41558-019-0398-8.

Ishimoto, Yuki, Masahiro Sugiyama, Etsushi Kato, Ryo Moriyama, Kazuhiro Tsuzuki, and Atsushi Kurosawa. 2017. "Putting Costs of Direct Air Capture in Context." SSRN Scholarly Paper ID 2982422. Rochester, NY: Social Science Research Network, 2017. https://papers.ssrn.com/abstract=2982422.

IUCN. 2015. "Ursus Maritimus: Ø. Wiig, S. Amstrup, T. Atwood, K. Laidre, N. Lunn, M. Obbard, E. Regehr, and G. Thiemann: The IUCN Red List of Threatened Species 2015: E.T22823A14871490." International Union for Conservation of Nature. 2015. https://doi.org/10.2305/IUCN.UK.2015-4.RLTS.T22823A14871490.en.

IUCN. 1986. *Polar Bears: Proceedings of the Ninth Working Meeting of the IUCN/SSC Polar Bear Specialist Group, Held at Edmonton, Canada, 9-11 August 1985*. IUCN Conservation Library. Cambridge, England: International Union for Conservation of Nature and Natural Resources, Conservation Monitoring Centre. http://pbsg.npolar.no/export/sites/pbsg/en/docs/PBSG09proc.pdf.

IUCN Polar Bear Specialist Group, and Working Meeting, eds. 1985. *Polar Bears: Proceedings of the Eighth Working Meeting of the IUCN/SSC Polar Bear Specialist Group, Held at the Ministry of Environment, Oslo, Norway, 15–19 January 1981*. Gland, Switzerland: International Union for Conservation of Nature and Natural Resources, 1985.

IUCN/SSC Polar Bear Specialist Group. 2019. "Status Report on the World's Polar Bear Subpopulations." 2019. http://pbsg.npolar.no/export/sites/pbsg/en/docs/2019-StatusReport.pdf.

Ivanic, Maros, Will Martin, and Hassan Zaman. 2011. "Estimating the Short-Run Poverty Impacts of the 2010–11 Surge in Food Prices." 2011. http://documents.worldbank.org/curated/en/560951468330321207/pdf/WPS5633.pdf.

Jakobsson, Martin, Antony Long, Ólafur Ingólfsson, Kurt H. Kjær, and Robert F. Spielhagen. 2010. "New Insights on Arctic Quaternary Climate Variability from

Palaeo-Records and Numerical Modelling." *Quaternary Science Reviews* 29, no. 25 (2010): 3349–58. https://doi.org/10.1016/j.quascirev.2010.08.016.

Jenkins, Jesse D. 2014. "Political Economy Constraints on Carbon Pricing Policies: What Are the Implications for Economic Efficiency, Environmental Efficacy, and Climate Policy Design?" *Energy Policy* 69 (June 2014): 467–77. https://doi.org/10.1016/j.enpol.2014.02.003.

Jevrejeva, S., L. P. Jackson, A. Grinsted, D. Lincke, and B. Marzeion. 2018. "Flood Damage Costs under the Sea Level Rise with Warming of 1.5 °C and 2 °C." *Environmental Research Letters* 13, no. 7 (2018): 074014. https://doi.org/10.1088/1748-9326/aacc76.

Jevrejeva, S., J. C. Moore, A. Grinsted, A. P. Matthews, and G. Spada. 2014. "Trends and Acceleration in Global and Regional Sea Levels since 1807." *Global and Planetary Change* 113 (February 2014): 11–22. https://doi.org/10.1016/j.gloplacha.2013.12.004.

Ji, Shuguang, Christopher R. Cherry, Matthew J. Bechle, Ye Wu, and Julian D. Marshall. 2012. "Electric Vehicles in China: Emissions and Health Impacts." *Environmental Science & Technology* 46, no. 4 (2012): 2018–24. https://doi.org/10.1021/es202347q.

Jiang, Leiwen. 2014. "Internal Consistency of Demographic Assumptions in the Shared Socioeconomic Pathways." *Population and Environment* 35, no. 3 (2014): 261–85. https://doi.org/10.1007/s11111-014-0206-3.

Johnsen, Reid, Jacob LaRiviere, and Hendrik Wolff. 2019. "Fracking, Coal, and Air Quality." *Journal of the Association of Environmental and Resource Economists* 6, no. 5 (2019): 1001–37. https://doi.org/10.1086/704888.

Johnson, Becky. 2019. "Climate Change Could Lead to Food Shortages in the UK Say MPs," 2019. https://news.sky.com/story/climate-change-could-lead-to-food-shortages-in-the-uk-say-mps-11811785.

Johnson, Nathanael. 2018. "Next-Gen Nuclear Is Coming, If We Want It." *Grist* (blog). July 18, 2018. https://grist.org/article/next-gen-nuclear-is-coming-if-we-want-it/.

Johnson, Stanley. 2012. *UNEP: The First 40 Years: A Narrative*. Nairobi: United Nations Environment Programme, 2012.

Jongman, Brenden, Hessel C. Winsemius, Jeroen C. J. H. Aerts, Erin Coughlan de Perez, Maarten K. van Aalst, Wolfgang Kron, and Philip J. Ward. 2015. "Declining Vulnerability to River Floods and the Global Benefits of Adaptation." *Proceedings of the National Academy of Sciences* 112, no. 18 (2015): E2271–E2280. https://doi.org/10.1073/pnas.1414439112.

Kahouli, Sondès. 2020. "An Economic Approach to the Study of the Relationship between Housing Hazards and Health: The Case of Residential Fuel Poverty in France." *Energy Economics* 85 (January 2020): 104592. https://doi.org/10.1016/j.eneco.2019.104592.

Kaiser, Brooks A. 2013. "Bioeconomic Factors of Natural Resource Transitions: The US Sperm Whale Fishery of the 19th Century." 116/13. Working Papers. University of Southern Denmark, Department of Sociology, Environmental and Business Economics. 2013. https://ideas.repec.org/p/sdk/wpaper/116.html.

Kalmus, Peter. 2019. "When I Started Speaking out on Climate Breakdown as a Scientist, I Was Afraid of Being Labeled 'Alarmist.'" 2019. https://twitter.com/ClimateHuman/status/1189300540494229504.

Kaplan, Sarah, and Emily Guskin. 2019. "Most American Teens Are Frightened by Climate Change, Poll Finds, and about 1 in 4 Are Taking Action." 2019. https://www .washingtonpost.com/science/most-american-teens-are-frightened-by-climate -change-poll-finds-and-about-1-in-4-are-taking-action/2019/09/15/1936da1c -d639-11e9-9610-fb56c5522e1c_story.html.

Kashi, Bahman. 2020. "Off-Grid Rural Electrification in Africa." Forthcoming. Copenhagen Consensus. 2020.

Kaufman, Mark. 2018. "How a Quiet California Town Protects Itself against Today's Megafires." 2018. https://mashable.com/article/wildfire-california-montecito -resistance/.

Kaur, Nehmat. 2017. "Ahmedabad: Cool Roofs Initiative with 5th Heat Action Plan." 2017. https://www.nrdc.org/experts/nehmat-kaur/ahmedabad-cool-roofs-initia tive-5th-heat-action-plan.

KC, Samir, and Wolfgang Lutz. 2017. "The Human Core of the Shared Socioeconomic Pathways: Population Scenarios by Age, Sex and Level of Education for All Countries to 2100." *Global Environmental Change* 42 (2017): 181–92. https://doi .org/10.1016/j.gloenvcha.2014.06.004.

Keeney, Ralph L. 1990. "Mortality Risks Induced by Economic Expenditures." *Risk Analysis* 10, no. 1 (1990): 147–59. https://doi.org/10.1111/j.1539-6924.1990.tb01029.x.

Keith, David W., Geoffrey Holmes, David St. Angelo, and Kenton Heidel. 2018. "A Process for Capturing CO_2 from the Atmosphere." *Joule* 2, no. 8 (2018): 1573–94. https://doi.org/10.1016/j.joule.2018.05.006.

Kench, Paul S., Murray R. Ford, and Susan D. Owen. 2018. "Patterns of Island Change and Persistence Offer Alternate Adaptation Pathways for Atoll Nations." *Nature Communications* 9, no. 1 (2018): 605. https://doi.org/10.1038/s41467-018-02954-1.

Kennedy, Robert. 1968. "Remarks at the University of Kansas, March 18, 1968 | JFK Library." 1968. https://www.jfklibrary.org/learn/about-jfk/the-kennedy-family /robert-f-kennedy/robert-f-kennedy-speeches/remarks-at-the-university-of -kansas-march-18-1968.

Kenward, Alyson, Daniel Yawitz, Todd Sanford, and Regina Wang. 2014. "Summer in the City: Hot and Getting Hotter." 2014. http://assets.climatecentral.org/pdfs /UrbanHeatIsland.pdf.

Khandker, Shahidur. 2012. "The Welfare Impacts of Rural Electrification in Bangladesh." *Energy Journal* 33, no. 1 (2012). https://doi.org/10.5547/ISSN0195-6574-EJ -Vol33-No1-7.

King, Myron, Daniel Altdorff, Pengfei Li, Lakshman Galagedara, Joseph Holden, and Adrian Unc. 2018. "Northward Shift of the Agricultural Climate Zone under 21st-Century Global Climate Change." *Scientific Reports* 8, no. 1 (2018): 7904. https://doi.org/10.1038/s41598-018-26321-8.

Kintisch, Eli. 2017. "U.S. Should Pursue Controversial Geoengineering Research, Federal Scientists Say for First Time." 2017. https://www.sciencemag.org/news /2017/01/us-should-pursue-controversial-geoengineering-research-federal -scientists-say-first.

Kloster, Silvia, and Gitta Lasslop. 2017. "Historical and Future Fire Occurrence (1850 to 2100) Simulated in CMIP5 Earth System Models." *Global and Planetary Change* 150 (March 2017): 58–69. https://doi.org/10.1016/j.gloplacha.2016.12.017.

Klotzbach, Philip J., Steven G. Bowen, Roger Pielke, and Michael Bell. 2018. "Continental U.S. Hurricane Landfall Frequency and Associated Damage: Observations and Future Risks." *Bulletin of the American Meteorological Society* 99, no. 7 (2018): 1359–76. https://doi.org/10.1175/BAMS-D-17-0184.1.

Knopf, Brigitte, Yen-Heng Henry Chen, Enrica De Cian, Hannah Förster, Amit Kanudia, Ioanna Karkatsouli, Ilkka Keppo, Tiina Koljonen, Katja Schumacher, and Detlef P. van Vuuren. 2013. "Beyond 2020—Strategies and Costs for Transforming the European Energy System." *Climate Change Economics* 04, supp. 01 (2013): 1340001. https://doi.org/10.1142/S2010007813400010.

Knorr, W., T. Kaminski, A. Arneth, and U. Weber. 2014. "Impact of Human Population Density on Fire Frequency at the Global Scale." *Biogeosciences* 11, 4 (2014): 1085–1102. https://doi.org/10.5194/bg-11-1085-2014.

Kohler, Hans-Peter, and Jere Behrman. 2018. "Benefits and Costs of the Population and Demography Targets for the Post-2015 Development Agenda." In *Prioritizing Development*, edited by Bjorn Lomborg, Cambridge UK: Cambridge University Press, 2018, 375–94. https://www.copenhagenconsensus.com/publication/post-2015-consensus-population-and-demography-assessment-kohler-behrman.

Kolbert, Elizabeth. 2006. *Field Notes from a Catastrophe: Man, Nature, and Climate Change.* New York: Bloomsbury Publishing, 2006.

Kongsager, Rico. 2018. "Linking Climate Change Adaptation and Mitigation: A Review with Evidence from the Land-Use Sectors." *Land* 7, no. 4 (2018): 158. https://doi.org/10.3390/land7040158.

Kotchen, Matthew J., Zachary M. Turk, and Anthony A. Leiserowitz. 2017. "Public Willingness to Pay for a US Carbon Tax and Preferences for Spending the Revenue." *Environmental Research Letters* 12, no. 9 (2017): 094012. https://doi.org/10.1088/1748-9326/aa822a.

Koubi, Vally. 2019. "Climate Change and Conflict." *Annual Review of Political Science* 22, no. 1 (2019): 343–60. https://doi.org/10.1146/annurev-polisci-050317-070830.

Kriegler, Elmar, Gunnar Luderer, Nico Bauer, Lavinia Baumstark, Shinichiro Fujimori, Alexander Popp, Joeri Rogelj, Jessica Strefler, and Detlef P. van Vuuren. 2018. "Pathways Limiting Warming to 1.5°C: A Tale of Turning Around in No Time?" *Philosophical Transactions of the Royal Society A: Mathematical, Physical and Engineering Sciences* 376, no. 2119 (2018): 20160457. https://doi.org/10.1098/rsta.2016.0457.

Kulp, Scott A., and Benjamin H. Strauss. 2019. "New Elevation Data Triple Estimates of Global Vulnerability to Sea-Level Rise and Coastal Flooding." *Nature Communications* 10, no. 1 (2019): 4844. https://doi.org/10.1038/s41467-019-12808-z.

Kunzig, Robert. 2008. "A Sunshade for Planet Earth." *Scientific American* 299, no. 5 (2008): 46–55.

Lane, Lee, J. Eric Bickel, Isabel Galiana, Chris Green, and Valentina Bosetti. 2009. "Advice for Policymakers." 2009. https://www.copenhagenconsensus.com/sites/default/files/cop15_policy_advice.pdf.

Lang, Peter. 2017. "Nuclear Power Learning and Deployment Rates; Disruption and Global Benefits Forgone." *Energies* 10, no. 12 (2017): 2169. https://doi.org/10.3390/en10122169.

Lazard. 2019a. *Lazard's Levelized Cost of Energy Analysis—Version 13.0.* 2019. https://www.lazard.com/media/451086/lazards-levelized-cost-of-energy-version-130-vf.pdf.

———. 2019b. *Lazard's Levelized Cost of Storage Analysis—Version 5.0.* 2019. https://www.lazard.com/media/451087/lazards-levelized-cost-of-storage-version-50-vf.pdf.

Leahy, Eimear, Seán Lyons, and Richard S. J. Tol. 2010. "An Estimate of the Number of Vegetarians in the World." 2010. https://www.esri.ie/system/files/media/file-uploads/2015-07/WP340.pdf.

Lee, Harry F. 2018. "Measuring the Effect of Climate Change on Wars in History." *Asian Geographer* 35, no. 2 (2018): 123–42. https://doi.org/10.1080/10225706.2018.1504807.

Lee, Kenneth, Edward Miguel, and Catherine Wolfram. 2016. "Appliance Ownership and Aspirations among Electric Grid and Home Solar Households in Rural Kenya." *American Economic Review* 106, no. 5 (2016): 89–94. https://doi.org/10.1257/aer.p20161097.

Letta, Marco, Pierluigi Montalbano, and Richard S. J. Tol. 2018. "Temperature Shocks, Short-Term Growth and Poverty Thresholds: Evidence from Rural Tanzania." *World Development* 112 (December 2018): 13–32. https://doi.org/10.1016/j.worlddev.2018.07.013.

Li, Fang, David M. Lawrence, and Ben Bond-Lamberty. 2018. "Human Impacts on 20th Century Fire Dynamics and Implications for Global Carbon and Water Trajectories." *Global and Planetary Change* 162 (2018): 18–27. https://doi.org/10.1016/j.gloplacha.2018.01.002.

Li, J., M. Hamdi-Cherif, and C. Cassen. 2017. "Aligning Domestic Policies with International Coordination in a Post-Paris Global Climate Regime: A Case for China." *Technological Forecasting and Social Change* 125 (2017): 258–74. https://doi.org/10.1016/j.techfore.2017.06.027.

Life. 1970. "Ecology: A Cause Becomes a Mass Movement." January 30 1970, 22–30. https://books.google.dk/books?id=bFAEAAAAMBAJ&lpg=PA3&pg=PA22#v=onepage&q&f=false.

Lim, Wee Ho, Dai Yamazaki, Sujan Koirala, Yukiko Hirabayashi, et al. 2018. "Long-Term Changes in Global Socioeconomic Benefits of Flood Defenses and Residual Risk Based on CMIP5 Climate Models." *Earth's Future* 6, no. 7 (2018): 938–54. http://dx.doi.org.stanford.idm.oclc.org/10.1002/2017EF000671.

Lincke, Daniel, and Jochen Hinkel. 2018. "Economically Robust Protection against 21st Century Sea-Level Rise." *Global Environmental Change-Human and Policy Dimensions* 51 (July 2018): 67–73. https://doi.org/10.1016/j.gloenvcha.2018.05.003.

Lino, Mark. 2017. "The Cost of Raising a Child." 2017. https://www.usda.gov/media/blog/2017/01/13/cost-raising-child.

Little, Amanda. 2019. "Climate Change Is Likely to Devastate the Global Food Supply," 2019. https://time.com/5663621/climate-change-food-supply/.

Lo, Y. T. Eunice, Daniel M. Mitchell, Antonio Gasparrini, Ana M. Vicedo-Cabrera, Kristie L. Ebi, Peter C. Frumhoff, Richard J. Millar, et al. 2019. "Increasing Mitigation Ambition to Meet the Paris Agreement's Temperature Goal Avoids Substantial Heat-Related Mortality in U.S. Cities." *Science Advances* 5, no. 6 (2019): eaau4373. https://doi.org/10.1126/sciadv.aau4373.

Lomborg, Bjorn. 2001. *The Skeptical Environmentalist: Measuring the Real State of the World*. Cambridge, UK: Cambridge University Press, 2001.

———. ed. 2010. *Smart Solutions to Climate Change: Comparing Costs and Benefits*. Cambridge, UK: Cambridge University Press, 2010.

———. 2012. "Environmental Alarmism, Then and Now: The Club of Rome's Problem—and Ours." *Foreign Affairs* 91, no. 4 (2012): 24–40.

———. ed. 2013. *How Much Have Global Problems Cost the World? A Scorecard from 1900 to 2050*. Cambridge, UK: Cambridge University Press, 2013.

———. 2016. "Impact of Current Climate Proposals." *Global Policy* 7, no. 1 (2016): 109–18. https://doi.org/10.1111/1758-5899.12295.

———. ed. 2018. *Prioritizing Development: A Cost Benefit Analysis of the United Nations' Sustainable Development Goals*. New York: Cambridge University Press, 2018.

———. 2019. "Humans Can Survive Underwater." 2019. https://www.project-syndicate.org/commentary/rising-sea-levels-media-alarmism-by-bjorn-lomborg-2019-11.

———. 2020. "Welfare in the 21st Century: Increasing Development, Reducing Inequality, the Impact of Climate Change, and the Cost of Climate Policies." *Technological Forecasting and Social Change*. 156:119981. https://doi.org/10.1016/j.techfore.2020.119981.

Loomis, John, and Michelle Haefele. 2017. "Quantifying Market and Non-Market Benefits and Costs of Hydraulic Fracturing in the United States: A Summary of the Literature." *Ecological Economics* 138 (August 2017): 160–67. https://doi.org/10.1016/j.ecolecon.2017.03.036.

Lovins, Amory B. 1976. "Energy Strategy: The Road Not Taken?" *Foreign Affairs* 55, no. 1 (1976): 65–96. https://doi.org/10.2307/20039628.

Lu, Denise, and Christopher Flavelle. 2019. "Rising Seas Will Erase More Cities by 2050, New Research Shows." 2019. https://www.nytimes.com/interactive/2019/10/29/climate/coastal-cities-underwater.html.

Lusk, Jayson L., and F. Bailey Norwood. 2016. "Some Vegetarians Spend Less Money on Food, Others Don't." *Ecological Economics* 130 (October 2016): 232–42. https://doi.org/10.1016/j.ecolecon.2016.07.005.

Lutter, Randall, and John F. Morrall. 1994. "Health-Health Analysis: A New Way to Evaluate Health and Safety Regulation." *Journal of Risk and Uncertainty* 8, no. 1 (1994): 43–66. https://doi.org/10.1007/BF01064085.

Lutter, Randall, John F. Morrall, and W. Kip Viscusi. 1999. "The Cost-Per-Life-Saved Cutoff For Safety-Enhancing Regulation." *Economic Inquiry* 37, no. 4 (1999): 599–608. https://doi.org/10.1111/j.1465-7295.1999.tb01450.x.

Lutz, Wolfgang, William P. Butz, and Samir KC, eds. 2014. *World Population and Human Capital in the Twenty-First Century*. First edition. Oxford: Oxford University Press, 2014.

Mach, Katharine J., Caroline M. Kraan, W. Neil Adger, Halvard Buhaug, Marshall Burke, James D. Fearon, Christopher B. Field, et al. 2019. "Climate as a Risk Factor for Armed Conflict." *Nature* 571, no. 7764 (2019): 193–97. https://doi.org/10.1038/s41586-019-1300-6.

MacKay, David. 2010. *Sustainable Energy—without the Hot Air*. Reprinted. Cambridge: UIT Cambridge, 2110.

Maddison, Angus. 2006. *The World Economy; Volume 1: A Millennial Perspective, Volume 2: Historical Statistics*. Paris: Development Centre of the Organisation for Economic Co-Operation And Development, 2006.

Magill, Bobby. 2019. "Miami, Other Coastal Cities May Drown in 80 Years, Study Says." 2019. https://news.bloombergenvironment.com/environment-and-energy/miami-other-coastal-cities-may-drown-in-80-years-study-says.

Mann, Michael L., Peter Berck, Max A. Moritz, Enric Batllori, James G. Baldwin, Conor K. Gately, and D. Richard Cameron. 2014. "Modeling Residential Development in California from 2000 to 2050: Integrating Wildfire Risk, Wildland and Agricultural Encroachment." *Land Use Policy* 41 (2014): 438–52. https://doi.org/10.1016/j.landusepol.2014.06.020.

Mao, Jiafu, Aurélien Ribes, Binyan Yan, Xiaoying Shi, Peter E. Thornton, Roland Séférian, Philippe Ciais, et al. 2016. "Human-Induced Greening of the Northern Extratropical Land Surface." *Nature Climate Change* 6, no. 10 (2016): 959–63. https://doi.org/10.1038/nclimate3056.

Markanday, Ambika, Ibon Galarraga, and Anil Markandya. 2019. "A Critical Review of Cost-Benefit Analysis for Climate Change Adaptation in Cities." *Climate Change Economics* 10, no. 04 (2019): 1950014. https://doi.org/10.1142/S2010007819500143.

Markandya, Anil, and Paul Wilkinson. 2007. "Electricity Generation and Health." *Lancet* 370, no. 9591 (2007): 979–90. https://doi.org/10.1016/S0140-6736(07)61253-7.

Marlon, J. R., P. J. Bartlein, C. Carcaillet, D. G. Gavin, S. P. Harrison, P. E. Higuera, F. Joos, M. J. Power, and I. C. Prentice. 2008. "Climate and Human Influences on Global Biomass Burning over the Past Two Millennia." *Nature Geoscience* 1 (September 2008): 697.

Marlon, J. R., Patrick J. Bartlein, Daniel G. Gavin, Colin J. Long, R. Scott Anderson, Christy E. Briles, Kendrick J. Brown, et al. 2012. "Long-Term Perspective on Wildfires in the Western USA." *Proceedings of the National Academy of Sciences* 109, no. 9 (2012): E535–E543. https://doi.org/10.1073/pnas.1112839109.

Martinko, Katherine. 2014. "Cut Your Carbon Footprint in Half by Going Vegetarian." 2014. https://www.treehugger.com/green-food/cut-your-carbon-footprint-half-going-vegetarian.html.

Matthews, H. Damon, Susan Solomon, and Raymond Pierrehumbert. 2012. "Cumulative Carbon as a Policy Framework for Achieving Climate Stabilization." *Philosophical Transactions of the Royal Society A: Mathematical, Physical and Engineering Sciences* 370, no. 1974 (2012): 4365–79. https://doi.org/10.1098/rsta.2012.0064.

Maxwell, Sean L., Richard A. Fuller, Thomas M. Brooks, and James E. M. Watson. 2016. "Biodiversity: The Ravages of Guns, Nets and Bulldozers." *Nature* 536, no. 7615 (2016): 143–45. https://doi.org/10.1038/536143a.

McAneney, John, Benjamin Sandercock, Ryan Crompton, Thomas Mortlock, Rade Musulin, Roger Pielke Jr., and Andrew Gissing. 2019. "Normalised Insurance Losses from Australian Natural Disasters: 1966–2017." *Environmental Hazards* 18, no. 5 (April 2019): 414–33. https://doi.org/10.1080/17477891.2019.1609406.

McKibben, Bill. 2019a. "Climate Change Is Shrinking Planet, in the Scariest Possible Way." 2019. https://twitter.com/billmckibben/status/1189277270944423939.

———. 2019b. *Falter: Has the Human Game Begun to Play Itself Out?* New York: Henry Holt, 2019.

McLean, Elena V., Sharmistha Bagchi-Sen, John D. Atkinson, Julia Ravenscroft, Sharon Hewner, and Alexandra Schindel. 2019. "Country-Level Analysis of Household Fuel Transitions." *World Development* 114 (February 2019): 267–80. https://doi.org/10.1016/j.worlddev.2018.10.006.

McPhate, Mike. 2017. "California Today: A Plan to Cool Down L.A." 2017. https://www.nytimes.com/2017/07/07/us/california-today-cool-pavements-la.html.

Meadows, Donella H., and Club of Rome, eds. 1972. *The Limits to Growth: A Report for the Club of Rome's Project on the Predicament of Mankind.* New York: Universe Books, 1972.

Meinshausen, M., S. C. B. Raper, and T. M. L. Wigley. 2011. "Emulating Coupled Atmosphere-Ocean and Carbon Cycle Models with a Simpler Model, MAGICC6—Part 1: Model Description and Calibration." *Atmos. Chem. Phys.* 11, no. 4 (2011): 1417–56. https://doi.org/10.5194/acp-11-1417-2011.

Melek, Nida Çakır, Michael Plante, and Mine Yücel. 2019. *The U.S. Shale Oil Boom, the Oil Export Ban, and the Economy: A General Equilibrium Analysis.* Working Paper 1708. 2019. https://doi.org/10.24149/wp1708.

Mendelsohn, Robert, Kerry Emanuel, Shun Chonabayashi, and Laura Bakkensen. 2012. "The Impact of Climate Change on Global Tropical Cyclone Damage." *Nature Climate Change* 2 (January 2012): 205.

Mexico NDC. 2016. "Mexico NDC Submission." 2016. https://www4.unfccc.int/sites/ndcstaging/PublishedDocuments/Mexico%20First/MEXICO%20INDC%2003.30.2015.pdf.

Meyer, Robinson. 2018. "A Disappointing New Problem With Geo-Engineering." *Atlantic*, August 8, 2018. https://www.theatlantic.com/science/archive/2018/08/solar-geo-engineering-cant-save-the-worlds-crops/567017/.

Mims, Christopher. 2009. "'Albedo Yachts' and Marine Clouds: A Cure for Climate Change?" 2009. https://www.scientificamerican.com/article/albedo-yachts-and-marine-clouds/.

Ministry for the Environment (New Zealand). 2019a. "Latest Update on New Zealand's 2020 Net Position." 2019. https://www.mfe.govt.nz/climate-change/climate-change-and-government/emissions-reduction-targets/reporting-our-targets-0.

———. 2019b. "New Zealand's Interactive Emissions Tracker." https://emissionstracker.mfe.govt.nz/#NrAMBoEYF12TwCIByBTAL02wBM4eigDs2AHEltEA.

MIT. 2015. "Energy and Climate Outlook 2015." 2015. http://globalchange.mit.edu/research/publications/other/special/2015Outlook.

Monbiot, George. 2007. "The Western Appetite for Biofuels Is Causing Starvation in the Poor World." 2007. https://www.theguardian.com/commentisfree/2007/nov/06/comment.biofuels.

Montana State University. 2019. "Best Practices for Wildfire Adaptation and Resilience." 2019. https://www.sciencedaily.com/releases/2019/08/190820101458.htm.

Morgan, Dan. 2002. "Enron also Courted Democrats; Chairman Pushed Firm's Agenda with Clinton White House." *Washington Post*, January 2002. https://www.washingtonpost.com/ac2/wp-dyn/?contentId=A37287-2002Jan12&node=&notFound=true&pagename=article.

Morrall III, John F. 2003. "Saving Lives: A Review of the Record." *Journal of Risk and Uncertainty* 27, no. 3 (2003): 221–37. https://doi.org/10.1023/A:1025841209892.

Moss, Todd, and Jacob Kincer. 2018. "Infographic—Are We Learning the Right Energy Lesson from Mobile Phones? The Energy Iceberg Says No." 2018. https://www.energyforgrowth.org/blog/infographic-are-we-learning-the-right-energy-lesson-from-mobile-phones-the-energy-iceberg-says-no/.

Mouillot, Florent, and Christopher B. Field. 2005. "Fire History and the Global Carbon Budget: A 1°× 1° Fire History Reconstruction for the 20th Century." *Global Change Biology* 11, no. 3 (2005): 398–420. https://doi.org/10.1111/j.1365-2486.2005.00920.x.

Munich Re. 2019. *The Natural Disasters of 2018 in Figures*. 2019. https://www.munichre.com/topics-online/en/climate-change-and-natural-disasters/natural-disasters/the-natural-disasters-of-2018-in-figures.html.

———. 2020. "Tropical Cyclones Cause Highest Losses | Munich Re." 2020. https://www.munichre.com/topics-online/en/climate-change-and-natural-disasters/natural-disasters/natural-disasters-of-2019-in-figures-tropical-cyclones-cause-highest-losses.html.

Muñoz, Leslie A., Francisco Olivera, Matthew Giglio, and Philip Berke. 2018. "The Impact of Urbanization on the Streamflows and the 100-Year Floodplain Extent of the Sims Bayou in Houston, Texas." *International Journal of River Basin Management* 16, no. 1 (2018): 61–69. https://doi.org/10.1080/15715124.2017.1372447.

Murtaugh, Paul A., and Michael G. Schlax. 2009. "Reproduction and the Carbon Legacies of Individuals." *Global Environmental Change* 19, no. 1 (2009): 14–20. https://doi.org/10.1016/j.gloenvcha.2008.10.007.

Nachmany, Michal, and Emily Mangan. 2018. *Aligning National and International Climate Targets*. 2018. http://www.lse.ac.uk/GranthamInstitute/wp-content/uploads/2018/10/Aligning-national-and-international-climate-targets-1.pdf.

Narassimhan, Easwaran, Kelly S. Gallagher, Stefan Koester, and Julio Rivera Alejo. 2018. "Carbon Pricing in Practice: A Review of Existing Emissions Trading Systems." *Climate Policy* 18, no. 8 (2018): 967–91. https://doi.org/10.1080/14693062.2018.1467827.

NAS. 2017. *A Century of Wildland Fire Research: Contributions to Long-Term Approaches for Wildland Fire Management: Proceedings of a Workshop*. Washington, DC: National Academies Press, 2017.

NASA. 2016. "Carbon Dioxide Fertilization Greening Earth, Study Finds." 2016. https://www.nasa.gov/feature/goddard/2016/carbon-dioxide-fertilization-greening-earth.

National Safety Council. 2019. "Vehicle Deaths Estimated at 40,000 for Third Straight Year." 2019. https://www.nsc.org/road-safety/safety-topics/fatality-estimates.

NCEI. 2019. *Billion-Dollar Weather and Climate Disasters.* 2019. https://www.ncdc.noaa.gov/billions/time-series.

NDRC. 2008. *Move Over, Gasoline: Here Come Biofuels.* 2008. https://web.archive.org/web/20081105142712/http://www.nrdc.org/air/transportation/biofuels.asp.

———. 2009. *Homegrown Energy from Biofuels: Fuel and Electricity Produced from Grass, Wood and Other Biomass Have the Potential to Help Repower America If Carefully Developed.* 2009. https://web.archive.org/web/20090805002845/http://www.nrdc.org/energy/biofuels.asp.

NDRRMC. 2014. "Situational Report Re Effects of Typhoon Yolanda (Haiyan)." 2014. http://www.ndrrmc.gov.ph/index.php/21-disaster-events/1329-situational-report-re-effects-of-typhoon-yolanda-haiyan.

Nebehay, Stephanie. 2008. "Top U.N. Human Rights Forum to Examine Food Crisis." 2008. https://www.reuters.com/article/us-un-food-rights/top-u-n-human-rights-forum-to-examine-food-crisis-idUSL0911964320080509.

NEED. 2019. *National Energy Efficiency Data-Framework (NEED): Consumption Data Tables 2019.* 2019. https://assets.publishing.service.gov.uk/government/uploads/system/uploads/attachment_data/file/812400/Consumption_headline_tables_2017.xlsx.

Nesbit, Jeffrey Asher. 2019. *This Is the Way the World Ends: How Droughts and Die-Offs, Heat Waves and Hurricanes Are Converging on America.* New York: Picador, 2019.

New York Times. 1970. "The End of Civilization Feared by Biochemist," November 19, 1970.

Newsweek. 1973. "Running out of Everything." 1973. http://www.flickr.com/photos/bob_roddis/5665427476/.

Nguyen, Van Kien, Jamie Pittock, and Daniel Connell. 2019. "Dikes, Rice, and Fish: How Rapid Changes in Land Use and Hydrology Have Transformed Agriculture and Subsistence Living in the Mekong Delta." *Regional Environmental Change* 19, no. 7 (2019): 2069–77. https://doi.org/10.1007/s10113-019-01548-x.

NIBS. 2018. "Natural Hazard Mitigation Saves Documents—National Institute of Building Sciences." 2018. https://www.nibs.org/page/ms2_dwnload.

Nicholls, Robert J., Natasha Marinova, Jason A. Lowe, Sally Brown, Pier Vellinga, Diogo de Gusmão, Jochen Hinkel, and Richard S. J. Tol. 2011. "Sea-Level Rise and Its Possible Impacts Given a 'beyond 4°C World' in the Twenty-First Century." *Philosophical Transactions of the Royal Society A: Mathematical, Physical and Engineering Sciences* 369, no. 1934 (2011): 161–81. https://doi.org/10.1098/rsta.2010.0291.

NIFC. 2019. *Total Wildland Fires and Acres by National Interagency Fire Center.* 2019. https://www.nifc.gov/fireInfo/fireInfo_stats_totalFires.html.

NOAA. 2019. "U.S. Percentage Areas (Very Warm/Cold, Very Wet/Dry)." 2019. ncdc.noaa.gov/temp-and-precip/uspa/wet-dry/0.

Nordhaus, Ted, and Alex Trembath. 2019. "Is Climate Change Like Diabetes or an Asteroid?" Breakthrough Institute. 2019. https://thebreakthrough.org/articles/is-climate-change-like-diabetes.

Nordhaus, William. 1975. "Can We Control Carbon Dioxide?" Monograph. June 1975. http://pure.iiasa.ac.at/id/eprint/365/.

———. 1991. "To Slow or Not to Slow: The Economics of The Greenhouse Effect." *Economic Journal* 101, no. 407 (1991): 920–37. https://doi.org/10.2307/2233864.

———. 2010. "Economic Aspects of Global Warming in a Post-Copenhagen Environment." *Proceedings of the National Academy of Sciences* 107, no. 26 (2010): 11721–26. https://doi.org/10.1073/pnas.1005985107.

———. 2013. *The Climate Casino: Risk, Uncertainty, and Economics for a Warming World*. New Haven: Yale University Press, 2013.

———. 2018. "Projections and Uncertainties about Climate Change in an Era of Minimal Climate Policies." *American Economic Journal: Economic Policy* 10, no. 3 (2018): 333–60. https://doi.org/10.1257/pol.20170046.

———. 2019a. "Economics of the Disintegration of the Greenland Ice Sheet." *Proceedings of the National Academy of Sciences* 116, no. 25 (2019): 12261–69. https://doi.org/10.1073/pnas.1814990116.

———. 2019b. "Can We Control Carbon Dioxide? (from 1975)." *American Economic Review* 109, no. 6 (2019): 2015–35. https://doi.org/10.1257/aer.109.6.2015.

Nordhaus, William, and Andrew Moffat. 2017. "A Survey of Global Impacts of Climate Change: Replication, Survey Methods, and a Statistical Analysis." w23646. Cambridge, MA: National Bureau of Economic Research. https://doi.org/10.3386/w23646.

Nordhaus, William, and Paul Sztorc. 2013. *DICE 2013R: Introduction and User's Manual*. 2013. http://www.econ.yale.edu/~nordhaus/homepage/homepage/documents/DICE_Manual_100413r1.pdf.

Norton, Michael, Andras Baldi, Vicas Buda, Bruno Carli, Pavel Cudlin, Mike B. Jones, Atte Korhola, et al. 2019. "Serious Mismatches Continue between Science and Policy in Forest Bioenergy." *GCB Bioenergy* 11, no. 11 (2019): 1256–63. https://doi.org/10.1111/gcbb.12643.

NRC. 2015. *Climate Intervention: Reflecting Sunlight to Cool Earth*. Washington, DC: National Academies Press, 2015. https://doi.org/10.17226/18988.

NZ Treasury. 2019. *Financial Statements of the Government of New Zealand for the Year Ended 30 June 2019*. 2019. https://treasury.govt.nz/publications/year-end/financial-statements-2019.

NZIER. 2018. "Economic Impact Analysis of 2050 Emissions Targets—A Dynamic Computable General Equilibrium Analysis." 2018. https://www.mfe.govt.nz/sites/default/files/media/Climate%20Change/NZIER%20report%20-%20Economic%20impact%20analysis%20of%202050%20emissions%20targets%20-%20FINAL.pdf.

OECD. 2019. *OECD Development Aid*. 2019. https://data.oecd.org/chart/5Mzc.

OECD-DAC. 2019. *Climate-Related Development Finance A Bilateral Provider Perspective*. 2019. https://public.tableau.com/views/Climate-relateddevelopmentfinance/CRDF-Donor.

Office for National Statistics. 2015a. "Monthly Provisional Figures on Deaths Registered in England and Wales." 2015. http://www.ons.gov.uk/ons/rel/vsob2/monthly

-figures-on-deaths-registered-by-area-of-usual-residence--england-and-wales
/april-2015--provisional-/rft-monthly-deaths-april-2015.xls.

———. 2015b. "Weekly Provisional Figures on Deaths Registered in England and Wales." 2015. http://www.ons.gov.uk/ons/rel/vsob2/weekly-provisional-figures -on-deaths-registered-in-england-and-wales/week-ending-13-02-2015/weekly -deaths—week-07-2015.xls.

———. 2017. "Household Disposable Income and Inequality in the UK: Financial Year Ending 2016." 2017. https://www.ons.gov.uk/file?uri=/peoplepopulation andcommunity/personalandhouseholdfinances/incomeandwealth/bulletins /householddisposableincomeandinequality/financialyearending2016/977e7b6f .xls.

Ofgem. 2018. "Energy Spend as a Percentage of Total Household Expenditure (UK)." 2018. https://www.ofgem.gov.uk/data-portal/energy-spend-percentage-total -household-expenditure-uk.

Olivié, D. J. L., D. Cariolle, H. Teyssèdre, D. Salas, A. Voldoire, H. Clark, D. Saint-Martin, et al. 2012. "Modeling the Climate Impact of Road Transport, Maritime Shipping and Aviation over the Period 1860–2100 with an AOGCM." *Atmospheric Chemistry and Physics* 12, no. 3 (2012): 1449–1480. https://doi.org/10.5194/acp-12-1449 -2012.

O'Neill, Brian C., Elmar Kriegler, Keywan Riahi, Kristie L. Ebi, Stephane Hallegatte, Timothy R. Carter, Ritu Mathur, and Detlef P. van Vuuren. 2014. "A New Scenario Framework for Climate Change Research: The Concept of Shared Socioeconomic Pathways." *Climatic Change* 122, no. 3 (2014): 387–400. https://doi.org/10.1007 /s10584-013-0905-2.

Oreskes, Naomi. 2015. "Playing Dumb on Climate Change," 2015. https://www.ny times.com/2015/01/04/opinion/sunday/playing-dumb-on-climate-change.html ?smid=tw-share&_r=1.

Oreskes, Naomi, and Nicholas Stern. 2019. "Climate Change Will Cost Us Even More Than We Think." 2019. https://www.nytimes.com/2019/10/23/opinion/climate -change-costs.html.

Ortiz-Ospina, Esteban. 2017. "The Size of the Poverty Gap: Some Hints Regarding the Cost of Ending Extreme Poverty." 2017. https://ourworldindata.org/size-poverty -gap.

Ortiz-Ospina, Esteban, and Max Roser. 2019. "Our World in Data: Child Labor." 2019. https://ourworldindata.org/child-labor.

Ostrander, Madeline. 2016. "How Do You Decide to Have a Baby When Climate Change Is Remaking Life on Earth?" 2016. https://www.thenation.com/article/how-do -you-decide-to-have-a-baby-when-climate-change-is-remaking-life-on-earth/.

Oxford University. 2011. "Lab-Grown Meat Would 'Cut Emissions and Save Energy.'" 2011. http://www.ox.ac.uk/news/2011-06-21-lab-grown-meat-would-cut -emissions-and-save-energy.

Page, Lucy, and Rohini Pande. 2018. "Ending Global Poverty: Why Money Isn't Enough." *Journal of Economic Perspectives* 32, no. 4 (2018): 173–99. https://doi.org/10.1257 /jep.32.4.173.

Palmissano, John. 1997. "Implications of the Climate Change Agreement in Kyoto & What Transpired." 1997. https://web.archive.org/web/20020918131105/http://www.cnsnews.com/specialreports/2002/kyoto_memo.asp.

Panos, Evangelos, and Martin Densing. 2019. "The Future Developments of the Electricity Prices in View of the Implementation of the Paris Agreements: Will the Current Trends Prevail, or a Reversal Is Ahead?" *Energy Economics*, 104476 (2019). https://doi.org/10.1016/j.eneco.2019.104476.

Parrish, David, and William Stockwell. 2015. "Urbanization and Air Pollution: Then and Now." *Eos* 96 (2015). doi:10.1029/2015EO021803.

Pascus, Brian. 2019. "Human Civilization Faces 'Existential Risk' by 2050 According to New Australian Climate Change Report." 2019. https://www.cbsnews.com/news/new-climate-change-report-human-civilization-at-risk-extinction-by-2050-new-australian-climate/.

Paul, Bimal Kanti. 2009. "Why Relatively Fewer People Died? The Case of Bangladesh's Cyclone Sidr." *Natural Hazards* 50, no. 2 (2009): 289–304. https://doi.org/10.1007/s11069-008-9340-5.

Paulson, Tom. 2013. "Epidemiology: A Mortal Foe." *Nature* 502, no. 7470 (2013): S2–3. https://doi.org/10.1038/502S2a.

Peterson, Thomas C., William M. Connolley, and John Fleck. 2008. "The Myth of the 1970s Global Cooling Scientific Consensus." *Bulletin of the American Meteorological Society* 89, no. 9 (2008): 1325–38. https://doi.org/10.1175/2008BAMS2370.1.

Pew Research Center. 2019. "Public's 2019 Priorities: Economy, Health Care, Education and Security All Near Top of List." 2019. https://www.people-press.org/2019/01/24/publics-2019-priorities-economy-health-care-education-and-security-all-near-top-of-list/.

Pielke, R. A., and C. W. Landsea. 1998. "Normalized Hurricane Damages in the United States: 1925–95." *Weather and Forecasting* 13, no. 3 (1998): 621–31.

Pielke, Roger. 2019. "Tracking Progress on the Economic Costs of Disasters under the Indicators of the Sustainable Development Goals." *Environmental Hazards* 18, no. 1 (2019): 1–6. https://doi.org/10.1080/17477891.2018.1540343.

Pielke, Roger A. 2009. "An Idealized Assessment of the Economics of Air Capture of Carbon Dioxide in Mitigation Policy." *Environmental Science & Policy* 12, no. 3 (2009): 216–25. https://doi.org/10.1016/j.envsci.2009.01.002.

Pires, J.C.M. 2019. "Negative Emissions Technologies: A Complementary Solution for Climate Change Mitigation." *Science of The Total Environment* 672 (July 2019): 502–14. https://doi.org/10.1016/j.scitotenv.2019.04.004.

Proctor, Jonathan, Solomon Hsiang, Jennifer Burney, Marshall Burke, and Wolfram Schlenker. 2018. "Estimating Global Agricultural Effects of Geoengineering Using Volcanic Eruptions." *Nature* 560, no. 7719 (2018): 480–83. https://doi.org/10.1038/s41586-018-0417-3.

Qiu, Linda. 2015. "Fact-Checking Bernie Sanders' Comments on Climate Change and Terrorism." 2015. https://www.politifact.com/truth-o-meter/statements/2015/nov/16/bernie-s/fact-checking-bernie-sanders-comments-climate-chan/.

Raimi, Daniel, and Gloria Aldana. 2018. "Understanding a New Study on Oil and Gas Methane Emissions." *Resources*, 2018. https://www.resourcesmag.org

/common-resources/understanding-a-new-study-on-oil-and-gas-methane
-emissions/.

Rao, Narasimha D., Petra Sauer, Matthew Gidden, and Keywan Riahi. 2019. "Income Inequality Projections for the Shared Socioeconomic Pathways (SSPs)." *Futures* 105 (2019): 27–39. https://doi.org/10.1016/j.futures.2018.07.001.

Rathi, Akshat. 2018. "Stacking Concrete Blocks Is a Surprisingly Efficient Way to Store Energy." *Quartz*. 2018. https://qz.com/1355672/stacking-concrete-blocks-is-a-surprisingly-efficient-way-to-store-energy/.

Ren, Xiaolin, Matthias Weitzel, Brian C. O'Neill, Peter Lawrence, Prasanth Meiyappan, Samuel Levis, Edward J. Balistreri, and Michael Dalton. 2018. "Avoided Economic Impacts of Climate Change on Agriculture: Integrating a Land Surface Model (CLM) with a Global Economic Model (IPETS)." *Climatic Change* 146, no. 3–4 (2018): 517–31. https://doi.org/10.1007/s10584-016-1791-1.

Retro Report. 2019. "Population Bomb: The Overpopulation Theory That Fell Flat." 2019. https://www.retroreport.org/video/the-population-bomb/.

Reuters. 2017. "China National Nuclear, Shenhua Team Up to Develop Gen-4 Reactor." Reuters. September 27, 2017. https://www.reuters.com/article/us-china-nuclear-idUSKCN1C2088.

———. 2019. "No Apologies: Africans Say Their Need for Oil Cash Outweighs Climate Concerns." Reuters. November 8, 2019. https://www.reuters.com/article/us-africa-oil-climate-idUSKBN1XI16X.

Reynolds, Jesse L., and Gernot Wagner. 2019. "Highly Decentralized Solar Geoengineering." *Environmental Politics*, July 2019, 1–17. https://doi.org/10.1080/09644016.2019.1648169.

Reynolds, R. V., and A. H. Pierson. 1941. *The Saw Timber Resource of the United States, 1630–1930*. U.S. Forest Service, Division of Forest Economics. 1941. https://www.fs.fed.us/research/docs/rpa/pre-1989/1941%20SAWTIMBER%20RESOURCE%20OF%20THE%20US%201630-1930.pdf.

Riahi, Keywan, Detlef P. van Vuuren, Elmar Kriegler, Jae Edmonds, Brian C. O'Neill, Shinichiro Fujimori, Nico Bauer, et al. 2017. "The Shared Socioeconomic Pathways and Their Energy, Land Use, and Greenhouse Gas Emissions Implications: An Overview." *Global Environmental Change* 42 (January 2017): 153–68. https://doi.org/10.1016/j.gloenvcha.2016.05.009.

Rijke, Jeroen, Sebastiaan van Herk, Chris Zevenbergen, and Richard Ashley. 2012. "Room for the River: Delivering Integrated River Basin Management in the Netherlands." *International Journal of River Basin Management* 10, no. 4 (2012): 369–82. https://doi.org/10.1080/15715124.2012.739173.

Ritchie, Hannah, and Max Roser. 2019a. "CO_2 and Greenhouse Gas Emissions." *Our World in Data*. 2019. https://ourworldindata.org/co2-and-other-greenhouse-gas-emissions.

———. 2019b. "Our World in Data: Sanitation Access." 2019. https://ourworldindata.org/sanitation.

Ritchie, Justin, and Hadi Dowlatabadi. 2017. "Why Do Climate Change Scenarios Return to Coal?" *Energy* 140 (2017): 1276–91. https://doi.org/10.1016/j.energy.2017.08.083.

Roberts, David. 2019. "If We Still Exist." 2019. https://twitter.com/drvox/status/1166772758782414848.

Rojas, Rodrigo, Luc Feyen, and Paul Watkiss. 2013. "Climate Change and River Floods in the European Union: Socio-Economic Consequences and the Costs and Benefits of Adaptation." *Global Environmental Change* 23, no. 6 (2013): 1737–51. https://doi.org/10.1016/j.gloenvcha.2013.08.006.

Romer, Christina D., and David H. Romer. 2010. "The Macroeconomic Effects of Tax Changes: Estimates Based on a New Measure of Fiscal Shocks." *American Economic Review* 100, no. 3 (2010): 763–801. https://doi.org/10.1257/aer.100.3.763.

Roser, Max. 2019a. "Our World in Data: Life Expectancy." 2019. https://ourworldindata.org/grapher/life-expectancy?time=1770..2019&country=OWID_WRL.

———. 2019b. "Our World in Data: War and Peace." 2019. https://ourworldindata.org/war-and-peace.

Roser, Max, and Esteban Ortiz-Ospina. 2019a. "Our World in Data: Global Extreme Poverty." 2019. https://ourworldindata.org/extreme-poverty.

———. 2019b. "Our World in Data: Literacy." https://ourworldindata.org/literacy.

Roser, Max, and Hannah Ritchie. 2019a. "Food per Person." *Our World in Data*. 2019. https://ourworldindata.org/food-per-person.

———. 2019b. "Malaria." 2019. https://ourworldindata.org/malaria.

Roser, Max, Hannah Ritchie, and Esteban Ortiz-Ospina. 2019. *World Population Growth*. 2019. https://ourworldindata.org/world-population-growth.

Rosling, Hans. 2012. *HDI Surprisingly Similar to GDP/Capita*. 2012. https://www.gapminder.org/news/hdi-surprisingly-similar-to-gdpcapita/.

Rowlatt, Justin. 2007. "We Are All Ethical Men and Women Now." 2007. https://www.bbc.co.uk/blogs/newsnight/2007/04/we_are_all_ethical_men_and_women_now.html.

Roy, Avik. 2014. "A Village That Harnessed the Sun." 2014. https://www.dailypioneer.com/2014/columnists/a-village-that-harnessed-the-sun.html.

Sacks, Daniel W., Betsey Stevenson, and Justin Wolfers. 2012. "The New Stylized Facts about Income and Subjective Well-Being." *Emotion* 12, no. 6 (2012): 1181–87. https://doi.org/10.1037/a0029873.

Sandström, Vilma, Hugo Valin, Tamás Krisztin, Petr Havlík, Mario Herrero, and Thomas Kastner. 2018. "The Role of Trade in the Greenhouse Gas Footprints of EU Diets." *Global Food Security* 19 (December 2018): 48–55. https://doi.org/10.1016/j.gfs.2018.08.007.

Sanz-Pérez, Eloy S., Christopher R. Murdock, Stephanie A. Didas, and Christopher W. Jones. 2016. "Direct Capture of CO_2 from Ambient Air." *Chemical Reviews* 116, no. 19 (2016): 11840–76. https://doi.org/10.1021/acs.chemrev.6b00173.

Satija, Neena, Kiah Collier, and Al Shaw. 2016. "Boomtown, Flood Town." 2016. https://www.scientificamerican.com/article/boomtown-flood-town/.

Savage, Luiza Ch. 2019. "The U.S. Left a Hole in Leadership on Climate. China Is Filling It." 2019. https://www.politico.com/story/2019/08/15/climate-china-global-translations-1662345.

Schelling, Thomas C. 1992. "Some Economics of Global Warming." *The American Economic Review* 82, no. 1 (1992): 1–14.

Schlosser, C. Adam, Kenneth Strzepek, Xiang Gao, Charles Fant, Élodie Blanc, Sergey Paltsev, Henry Jacoby, John Reilly, and Arthur Gueneau. 2014. "The Future of Global Water Stress: An Integrated Assessment: Schlosser et al." *Earth's Future* 2, no. 8 (2014): 341–61. https://doi.org/10.1002/2014EF000238.

Schobert, Harold H. 2002. *Energy and Society: An Introduction*. New York: Taylor & Francis, 2002.

Schuerch, Mark, Tom Spencer, Stijn Temmerman, Matthew L. Kirwan, Claudia Wolff, Daniel Lincke, Chris J. McOwen, et al. 2018. "Future Response of Global Coastal Wetlands to Sea-Level Rise." *Nature* 561, no. 7722 (2018): 231–34. https://doi.org/10.1038/s41586-018-0476-5.

Selby, Jan. 2019. "Climate Change and the Syrian Civil War, Part II: The Jazira's Agrarian Crisis." *Geoforum* 101 (May 2019): 260–74. https://doi.org/10.1016/j.geoforum.2018.06.010.

Sengupta, Somini, and Nadja Popovich. 2019. "More Than 60 Countries Say They'll Zero Out Carbon Emissions. The Catch? They're Not the Big Emitters." 2019. https://www.nytimes.com/interactive/2019/09/25/climate/un-net-zero-emissions.html.

Seo, S. Niggol, and Robert Mendelsohn. 2008. "An Analysis of Crop Choice: Adapting to Climate Change in South American Farms." *Ecological Economics* 67, no. 1 (2008): 109–16. https://doi.org/10.1016/j.ecolecon.2007.12.007.

Sharma, Ashish, Conrad Wasko, and Dennis P. Lettenmaier. 2018. "If Precipitation Extremes Are Increasing, Why Aren't Floods?" *Water Resources Research* 54, no. 11 (2018): 8545–51. https://doi.org/10.1029/2018WR023749.

Sharma, Rajesh. 2018. "Health and Economic Growth: Evidence from Dynamic Panel Data of 143 Years." *PLOS One* 13, no. 10 (2018): e0204940–e0204940. https://doi.org/10.1371/journal.pone.0204940.

Shellenberger, Michael. 2019. "The Reason Renewables Can't Power Modern Civilization Is Because They Were Never Meant To." 2019. https://www.forbes.com/sites/michaelshellenberger/2019/05/06/the-reason-renewables-cant-power-modern-civilization-is-because-they-were-never-meant-to/#5b0849cdea2b.

Shields, Fiona. 2019. "Why We're Rethinking the Images We Use for Our Climate Journalism." 2019. https://www.theguardian.com/environment/2019/oct/18/guardian-climate-pledge-2019-images-pictures-guidelines.

SIPRI. 2019. "World Military Expenditure Grows to $1.8 Trillion in 2018 | SIPRI." 2019. https://www.sipri.org/media/press-release/2019/world-military-expenditure-grows-18-trillion-2018.

Sivaram, Varun, and Shayle Kann. 2016. "Solar Power Needs a More Ambitious Cost Target." *Nature Energy* 1, no. 4 (2016): 16036. https://doi.org/10.1038/nenergy.2016.36.

Smil, Vaclav. 2014. "The Long, Slow Rise of Solar and Wind: The Great Hope for a Quick and Sweeping Transition to Renewable Energy Is Wishful Thinking." *Scientific American* 310, no. 12014 (January 2014), 52–57.

Smith, David. 2019. "'The Poor Are Punished': Dorian Lays Bare Inequality in the Bahamas." 2019. https://www.theguardian.com/world/2019/sep/13/hurricane-dorian-the-mudd-haitians-inequality.

Smith, Joel. 2019. "There Is Harm in Exaggerating Climate Change Risks." 2019. https://www.dailycamera.com/2019/08/24/joel-smith-harm-exaggerating-climate-change-risks/.

Smith, Matthew. 2019. "International Poll: Most Expect to Feel Impact of Climate Change, Many Think It Will Make Us Extinct." 2019. https://yougov.co.uk/topics/science/articles-reports/2019/09/15/international-poll-most-expect-feel-impact-climate.

Smyth, Jamie. 2019. "Climate Change Is Coming for Australian Sheep." 2019. https://www.ozy.com/around-the-world/climate-change-is-coming-for-australian-sheep/96774/.

Spratt, David, and Ian Dunlop. 2019. "Existential Climate-Related Security Risk: A Scenario Approach." 2019. https://docs.wixstatic.com/ugd/148cb0_a1406e0143ac4c469196d3003bc1e687.pdf.

St. John, Jeff. 2019. "Global Energy Storage to Hit 158 Gigawatt-Hours by 2024, Led by US and China." October 4, 2019. https://www.greentechmedia.com/articles/read/global-energy-storage-to-hit-158-gigawatt-hours-by-2024-with-u-s-and-china.

Statista. 2019. "Principal Rice Exporting Countries Worldwide in 2018/2019." 2019. https://www.statista.com/statistics/255947/top-rice-exporting-countries-worldwide-2011/.

Steckel, Jan Christoph, Narasimha D. Rao, and Michael Jakob. 2017. "Access to Infrastructure Services: Global Trends and Drivers." *Utilities Policy* 45 (April 2017): 109–17. https://doi.org/10.1016/j.jup.2017.03.001.

Sterman, John D, Lori Siegel, and Juliette N Rooney-Varga. 2018. "Does Replacing Coal with Wood Lower CO_2 Emissions? Dynamic Lifecycle Analysis of Wood Bioenergy." *Environmental Research Letters* 13, no. 1 (2018): 015007. https://doi.org/10.1088/1748-9326/aaa512.

Stevenson, Betsey, and Justin Wolfers. 2013. "Subjective Well-Being and Income: Is There Any Evidence of Satiation?" *American Economic Review* 103, no. 3 (2013): 598–604. https://doi.org/10.1257/aer.103.3.598.

Stieb, Matt. 2019. "Meeting Paris Climate Goals Would Save Thousands of American Lives During Heat Waves: Study." 2019. http://nymag.com/intelligencer/2019/06/paris-goals-would-save-thousands-of-u-s-lives-in-heat-waves.html.

Stiglitz, Joseph E., Jean-Paul Fitoussi, and Martine Durand. 2018. *Beyond GDP: Measuring What Counts for Economic and Social Performance*. Paris: OECD Publishing, 2018.

Stocker, Thomas F., and Intergovernmental Panel on Climate Change, eds. 2013. *Climate Change 2013: The Physical Science Basis; Summary for Policymakers, a Report of Working Group I of the IPCC, Technical Summary, a Report Accepted by Working Group I of the IPCC but Not Approved in Detail and Frequently Asked Questions; Part of the Working Group I Contribution to the Fifth Assessment Report of the Intergovernmental Panel on Climate Change*. New York: Intergovernmental Panel on Climate Change, 2013.

Strader, Stephen M. 2018. "Spatiotemporal Changes in Conterminous US Wildfire

Exposure from 1940 to 2010." *Natural Hazards* 92, no. 1 (2018): 543–65. https://doi .org/10.1007/s11069-018-3217-z.

Strader, Stephen M., and Walker S. Ashley. 2015. "The Expanding Bull's-Eye Effect." *Weatherwise* 68, no. 5 (2015): 23–29. https://doi.org/10.1080/00431672.2015.1067108.

Strife, Susan Jean. 2012. "Children's Environmental Concerns: Expressing Ecophobia." *Journal of Environmental Education* 43, no. 1 (2012): 37–54. https://doi.org/10.108 0/00958964.2011.602131.

Sullivan, Arthur. 2018. "To Fly or Not to Fly? The Environmental Cost of Air Travel." 2018. https://www.dw.com/en/to-fly-or-not-to-fly-the-environmental-cost-of -air-travel/a-42090155.

Syphard, Alexandra D., Jon E. Keeley, Anne H. Pfaff, and Ken Ferschweiler. 2017. "Human Presence Diminishes the Importance of Climate in Driving Fire Activity across the United States." *Proceedings of the National Academy of Sciences* 114, no. 52 (2017): 13750–55. https://doi.org/10.1073/pnas.1713885114.

Tengs, Tammy O. 1996. "Dying Too Soon: How Cost-Effectiveness Analysis Can Save Lives." 1996. http://www.ncpathinktank.org/pdfs/st204.pdf.

Tengs, Tammy O., M. E. Adams, J. S, Pliskin, D. G, Safran, J. E. Siegel, M. C. Weinstein, and J. D. Graham. 1995. "500 Lifesaving Interventions and Their Cost-Effectiveness." *Risk Analysis* 15, no. 3 (1995): 369–90.

Tengs, Tammy O., and John D. Graham. 1996. "The Opportunity Costs of Haphazard Social Investements in Life-Saving." In *Risks, Costs, and Lives Saved: Getting Better Results from Regulation*, edited by Robert W. Hahn. Oxford: Oxford University Press,1996.

TerraPower. 2019. "TWR Technology: Preparing Nuclear Energy for Global Growth." 2019. https://terrapower.com/productservices/twr.

Terrenoire, E., D. A. Hauglustaine, T. Gasser, and O. Penanhoat. 2019. "The Contribution of Carbon Dioxide Emissions from the Aviation Sector to Future Climate Change." *Environmental Research Letters* 14, no. 8 (2019): 084019. https://doi .org/10.1088/1748-9326/ab3086.

Tessum, Christopher W., Jason D. Hill, and Julian D. Marshall. 2014. "Life Cycle Air Quality Impacts of Conventional and Alternative Light-Duty Transportation in the United States." *Proceedings of the National Academy of Sciences*. 2014. https:// doi.org/10.1073/pnas.1406853111.

Thompson, Megan. 2019. "As U.S. Cedes Leadership on Climate, China Steps Up." 2019. https://www.pbs.org/newshour/show/as-us-cedes-leadership-on-climate -china-steps-up.

Tidman, Zoe. 2019. "New Zealand Passes 'Zero Carbon' Law in Fight against Climate Change." 2019. https://www.independent.co.uk/news/world/australasia/new -zealand-zero-carbon-law-emissions-jacinda-ardern-climate-change-a9189341 .html.

Time. 1972. "Environment: The Worst Is Yet to Be?" *Time*, January 24, 1972, 32

———. 2006. Special Report Global Warming. 2006.

———. 2019. Our Sinking Planet. 2019. https://time.com/magazine/us/5606236/june -24th-2019-vol-193-no-24-u-s/.

Timmers, Victor R. J. H., and Peter A. J. Achten. 2016. "Non-Exhaust PM Emissions from Electric Vehicles." *Atmospheric Environment* 134 (2016): 10–17. https://doi.org/10.1016/j.atmosenv.2016.03.017.

Tol, Richard S. J. 2018. "The Economic Impacts of Climate Change." *Review of Environmental Economics and Policy* 12, no. 1 (2018): 4–25. https://doi.org/10.1093/reep/rex027.

———. 2019. *Climate Economics: Economic Analysis of Climate, Climate Change and Climate Policy.* Second edition. Cheltenham, UK: Edward Elgar Publishing, 2019.

Tol, Richard S. J., and Hadi Dowlatabadi. 2001. "Vector-Borne Diseases, Development & Climate Change." *Integrated Assessment* 2, no. 4 (2001): 173–81. https://doi.org/10.1023/A:1013390516078.

Trachtenberg, Marc. 2018. "Assessing Soviet Economic Performance during the Cold War: A Failure of Intelligence?" *Texas National Security Review* 1, no. 2 (2018). https://2llqix3cnhb21kcxpr2u9o1k-wpengine.netdna-ssl.com/wp-content/uploads/2018/04/TNSR-Journal-Issue-2-Tratchenberg.pdf.

Trivedi, Anjani. 2018. "The $6 Trillion Barrier Holding Electric Cars Back." 2018. https://www.bloomberg.com/opinion/articles/2018-11-04/electric-cars-face-a-6-trillion-barrier-to-widespread-adoption.

UNAIDS. 2019. *Global AIDS Update 2019—Communities at the Centre.* 2019. https://www.unaids.org/en/resources/documents/2019/2019-global-AIDS-update.

UNDESA. 2019. *UN World Population Prospects 2019.* 2019. https://population.un.org/wpp/DataQuery/.

UNEP. 1982. "The State of the Environment 1972–1982." 1982. https://wedocs.unep.org/handle/20.500.11822/28253.

———. 2014. *Progress towards the Aichi Biodiversity Targets: An Assessment of Biodiversity Trends, Policy Scenarios and Key Actions.* 2014. https://www.cbd.int/gbo4/.

———. 2019. *Lessons from a Decade of Emissions Gap Assessments.* 2019. https://wedocs.unep.org/bitstream/handle/20.500.11822/30022/EGR10.pdf?sequence=1&isAllowed=y.

UNFCCC. 2015. "Synthesis Report on the Aggregate Effect of the Intended Nationally Determined Contributions." 2015. http://unfccc.int/resource/docs/2015/cop21/eng/07.pdf.

———. 2018a. "The People's Seat—Transcript of the Speech by Sir David Attenborough." 2018. https://unfccc.int/sites/default/files/resource/The%20People%27s%20Address%202.11.18_FINAL.pdf.

———. 2018b. "What Is the Paris Agreement?" 2018. https://unfccc.int/process-and-meetings/the-paris-agreement/what-is-the-paris-agreement.

———. 2019. "António Guterres: Climate Change Is Biggest Threat to Global Economy." 2019. https://unfccc.int/news/antonio-guterres-climate-change-is-biggest-threat-to-global-economy.

United Nations. 2019a. "The World We Want." 2019. http://data.myworld2015.org/.

———. 2019b. "World Population Prospects 2019." 2019. https://population.un.org/wpp/DataQuery/.

United States Department of Agriculture Economic Research Service. 2019. "Ag and Food Sectors and the Economy." 2019. https://www.ers.usda.gov/data-products

/ag-and-food-statistics-charting-the-essentials/ag-and-food-sectors-and-the -economy/.

U.S. Global Change Research Program, D. J. Wuebbles, D. W. Fahey, K. A. Hibbard, D. J. Dokken, B. C. Stewart, and T. K. Maycock. 2017. "Climate Science Special Report: Fourth National Climate Assessment, Volume I." U.S. Global Change Research Program. 2017. https://doi.org/10.7930/J0J964J6.

USGCRP. 2017. *Climate Science Special Report: Fourth National Climate Assessment.* U.S. Global Change Research Program. Washington, DC. 2017. https://science2017. globalchange.gov/downloads/CSSR2017_FullReport.pdf.

———. 2018. *Fourth National Climate Assessment 2018: Impacts, Risks, and Adaptation in the United States.* U.S. Global Change Research Program. Washington, DC. 2018. https://science2017.globalchange.gov/downloads/CSSR2017_FullReport .pdf.

USNDC. 2016. "United States NDC Submission." 2016. https://www4.unfccc.int/sites /ndcstaging/PublishedDocuments/United%20States%20of%20America%20First /U.S.A.%20First%20NDC%20Submission.pdf.

Vaidyanathan, Gayathri. 2015. "Coal Trumps Solar in India." 2015. https://www.scien tificamerican.com/article/coal-trumps-solar-in-india/.

van Vuuren, Detlef P., Elmar Kriegler, Brian C. O'Neill, Kristie L. Ebi, Keywan Riahi, Timothy R. Carter, Jae Edmonds, et al. 2014. "A New Scenario Framework for Climate Change Research: Scenario Matrix Architecture." *Climatic Change* 122, no. 3 (2014): 373–86. https://doi.org/10.1007/s10584-013-0906-1.

Vassall, Anna. 2018. "Benefits and Costs of TB Control for the Post-2015 Development Agenda." In *Prioritizing Development*, edited by Bjorn Lomborg, Cambridge, UK: Cambridge University Press, 2018, 255–65 https://www.copenhagenconsensus .com/publication/post-2015-consensus-health-perspective-tuberculosis-vassall.

Verkaik, Robert. 2009. "Just 96 Months to Save World, Says Prince Charles." 2009. https://www.independent.co.uk/environment/green-living/just-96-months-to -save-world-says-prince-charles-1738049.html.

Veysey, Jason, Claudia Octaviano, Katherine Calvin, Sara Herreras Martinez, Alban Kitous, James McFarland, and Bob van der Zwaan. 2016. "Pathways to Mexico's Climate Change Mitigation Targets: A Multi-Model Analysis." *Energy Economics* 56 (2016): 587–99. https://doi.org/10.1016/j.eneco.2015.04.011.

Vicedo-Cabrera, Ana M., Francesco Sera, Yuming Guo, Yeonseung Chung, Katherine Arbuthnott, Shilu Tong, Aurelio Tobias, et al. 2018. "A Multi-Country Analysis on Potential Adaptive Mechanisms to Cold and Heat in a Changing Climate." *Environment International* 111 (2018): 239–46. https://doi.org/10.1016/j.envint.2017.11.006.

Victor, David G., Keigo Akimoto, Yoichi Kaya, Mitsutsune Yamaguchi, Danny Cullenward, and Cameron Hepburn. 2017. "Prove Paris Was More Than Paper Promises." *Nature* 548, no. 7665 (2017): 25–27. https://doi.org/10.1038/548025a.

Vidal, John. 2014. "Yeb Sano: Unlikely Climate Justice Star." 2014. https://www .theguardian.com/environment/2014/apr/01/yeb-sano-typhoon-haiyan-un -climate-talks.

Vidal, John, Adam Vaughan, Suzanne Goldenberg, Lenore Taylor, and Daniel Boffey. 2015. "World Leaders Hail Paris Climate Deal as 'Major Leap for Mankind.'" 2015.

https://www.theguardian.com/environment/2015/dec/13/world-leaders-hail-paris-climate-deal.

Wallace-Wells, David. 2017. "The Uninhabitable Earth." 2017. http://nymag.com/intelligencer/2017/07/climate-change-earth-too-hot-for-humans.html.

———. 2019a. *The Uninhabitable Earth.* New York: Tim Duggan Books, 2019.

———. 2019b. "We're Getting a Clearer Picture of the Climate Future—and It's Not as Bad as It Once Looked." *New York*, December 20, 2019. https://nymag.com/intelligencer/2019/12/climate-change-worst-case-scenario-now-looks-unrealistic.html.

Walter, David, ed. 1992. *Today Then: America's Best Minds Look 100 Years into the Future on the Occasion of the 1893 World's Columbian Exposition.* Helena, MT: American & World Geographic Pub., 1992.

Wang, Jianliang, Lianyong Feng, Xu Tang, Yongmei Bentley, and Mikael Höök. 2017. "The Implications of Fossil Fuel Supply Constraints on Climate Change Projections: A Supply-Side Analysis." *Futures* 86 (2017): 58–72. https://doi.org/10.1016/j.futures.2016.04.007.

Wanner, Brent. 2019. *Is Exponential Growth of Solar PV the Obvious Conclusion?* 2019. https://www.iea.org/commentaries/is-exponential-growth-of-solar-pv-the-obvious-conclusion.

Ward, Daniel S., Elena Shevliakova, Sergey Malyshev, and Sam Rabin. 2018. "Trends and Variability of Global Fire Emissions Due to Historical Anthropogenic Activities." *Global Biogeochemical Cycles* 32, no. 1 (2018): 122–42. https://doi.org/10.1002/2017GB005787.

Ward, Victoria. 2015. "Winter Death Toll 'to Exceed 40,000.'" 2015. https://www.telegraph.co.uk/news/weather/11382808/Winter-death-toll-to-exceed-40000.html.

Washington Herald. 1912. "15,000 Die in Philippine Storm." 1912. https://chroniclingamerica.loc.gov/lccn/sn83045433/1912-11-30/ed-1/seq-1/.

Washington Post. 2019. "Sea-Level Rise Could Be Even Worse than We've Been Led to Expect." 2019. https://www.washingtonpost.com/opinions/sea-level-rise-could-be-even-worse-than-weve-been-led-to-expect/2019/05/30/7eb5a7f8-7d9f-11e9-8ede-f4abf521ef17_story.html.

Watts, Jonathan. 2018. "We Have 12 Years to Limit Climate Change Catastrophe, Warns UN." 2018. https://www.theguardian.com/environment/2018/oct/08/global-warming-must-not-exceed-15c-warns-landmark-un-report.

Watts, Nick, Markus Amann, Nigel Arnell, Sonja Ayeb-Karlsson, Kristine Belesova, Helen Berry, Timothy Bouley, et al. 2018. "The 2018 Report of the Lancet Countdown on Health and Climate Change: Shaping the Health of Nations for Centuries to Come." *Lancet* 392, no. 10163 (2018): 2479–2514. https://doi.org/10.1016/S0140-6736(18)32594-7.

Weinkle, Jessica, Chris Landsea, Douglas Collins, Rade Musulin, Ryan P. Crompton, Philip J. Klotzbach, and Roger Pielke. 2018. "Normalized Hurricane Damage in the Continental United States 1900–2017." *Nature Sustainability* 1, no. 12 (2018): 808–13. https://doi.org/10.1038/s41893-018-0165-2.

Weise, Elizabeth. 2019. "End of Civilization: Climate Change Apocalypse Could Start by 2050 If We Don't Act, Report Warns." 2019. https://www.usatoday.com/story

/news/nation/2019/06/05/climate-change-apocalypse-could-start-2050-if-we
-do-noting/1356865001/.

Weiss, Thomas J. 1992. *U.S. Labor Force Estimates and Economic Growth, 1800–1860*.
NBER Book Chapter Series, no. c8007. Cambridge, MA: National Bureau of Economic Research, 1992. http://www.nber.org/papers/c8007.

WHO. 2006. *Fuel for Life. Household Energy and Health*. 2006. https://www.who.int
/airpollution/publications/fuelforlife.pdf.

———. 2018. "Family Planning/Contraception." 2018. https://www.who.int/news
-room/fact-sheets/detail/family-planning-contraception.

———. 2019. *Cancer Mortality Database*. 2019. http://www-dep.iarc.fr/WHOdb
/WHOdb.htm.

Wiig, Øystein, Erik W. Born, Gerald Warren Garner, and International Union for Conservation of Nature and Natural Resources, eds. 1995. *Polar Bears: Proceedings of
the Eleventh Working Meeting of the IUCN/SSC Polar Bear Specialist Group, 25–27
January 1993, Copenhagen, Denmark*. Occasional Paper of the IUCN Species Survival Commission (SSC) 10. Gland, Switzerland: IUCN, 1995.

Wilby, R. L., and G. L. W Perry. 2006. "Climate Change, Biodiversity and the Urban
Environment: A Critical Review Based on London, UK." *Progress in Physical Geography* 30, no. 1 (2006): 73–98.

Williams, Rebecca Jane. 2014. "Storming the Citadels of Poverty: Family Planning under the Emergency in India, 1975–1977." *Journal of Asian Studies* 73, no. 2 (2014):
471–92. https://doi.org/10.1017/S0021911813002350.

Williams, Timothy. 2019. "Why Oregon Republicans Vanished Over a Climate Change
Vote." *New York Times*, sec. US, June 24, 2019. https://www.nytimes.com/2019
/06/24/us/oregon-climate-change-walkout.html.

Willsher, Kim. 2018. "Macron Scraps Fuel Tax Rise in Face of Gilets Jaunes Protests."
2018. https://www.theguardian.com/world/2018/dec/05/france-wealth-tax
-changes-gilets-jaunes-protests-president-macron.

Woerdman, E., O. Couwenberg, and A. Nentjes. 2009. "Energy Prices and Emissions Trading: Windfall Profits from Grandfathering?" *European Journal of Law
and Economics* 28, no. 2 (2009): 185–202. https://doi.org/10.1007/s10657-009
-9098-6.

World Bank. 2017. "Bangladesh Continues to Reduce Poverty But at Slower Pace."
2017. https://www.worldbank.org/en/news/feature/2017/10/24/bangladesh
-continues-to-reduce-poverty-but-at-slower-pace.

———. 2019a. "Agriculture, Forestry, and Fishing, Value Added (% of GDP)—Low &
Middle Income, World." 2019. https://data.worldbank.org/indicator/NV.AGR
.TOTL.ZS?locations=XO-1W.

———. 2019b. "Employment in Agriculture (% of Total Employment) (Modeled ILO
Estimate)—Low & Middle Income, World." 2019. https://data.worldbank.org
/indicator/SL.AGR.EMPL.ZS?locations=XO-1W.

———. 2019c. "GDP per Capita, PPP (Constant 2011 International $)." 2019. https://
data.worldbank.org/indicator/NY.GDP.PCAP.PP.KD.

———. 2019d. "GNI per Capita, Atlas Method (Current US$)—China." 2019. https://
data.worldbank.org/indicator/NY.GNP.PCAP.CD?locations=CN.

———. 2019e. *World Development Indicators Online*. 2019. https://databank.world bank.org/data/reports.aspx?source=world-development-indicators.

World Health Organization, UNICEF, and WHO/UNICEF Joint Water Supply and Sanitation Monitoring Programme. 2015. *25 Years Progress on Sanitation and Drinking Water: 2015 Update and MDG Assessment*. 2015.

World Resources Institute. 2019a. "CAIT Climate Data Explorer—UNFCCC Annex I." http://cait.wri.org/profile/UNFCCC%20Annex%20I. 2019.

———. 2019b. "CAIT Climate Data Explorer—United Kingdom." http://cait.wri.org /profile/United%20Kingdom.

WWF. 2007. *WWF Position on Biofuels in the EU*. 2007. http://assets.panda.org/down loads/wwf_position_eu_biofuels.pdf.

———. 2018. "Living Planet Report—2018." Gland, Switzerland. 2018. https://c402277 .ssl.cf1.rackcdn.com/publications/1187/files/original/LPR2018_Full_Report_ Spreads.pdf.

Yang, Jia, Hanqin Tian, Bo Tao, Wei Ren, John Kush, Yongqiang Liu, and Yuhang Wang. 2014. "Spatial and Temporal Patterns of Global Burned Area in Response to Anthropogenic and Environmental Factors: Reconstructing Global Fire History for the 20th and Early 21st Centuries." *Journal of Geophysical Research: Biogeosciences* 119, no. 3 (2014): 249–63. https://doi.org/10.1002/2013JG002532.

Yang, Pu, Yun-Fei Yao, Zhifu Mi, Yun-Fei Cao, Hua Liao, Bi-Ying Yu, Qiao-Mei Liang, D'Maris Coffman, and Yi-Ming Wei. 2018. "Social Cost of Carbon under Shared Socioeconomic Pathways." *Global Environmental Change* 53 (November 2018): 225–32. https://doi.org/10.1016/j.gloenvcha.2018.10.001.

Young, David, and John Bistline. 2018. "The Costs and Value of Renewable Portfolio Standards in Meeting Decarbonization Goals." *Energy Economics* 73 (June 2018): 337–51. https://doi.org/10.1016/j.eneco.2018.04.017.

Zagorsky, Jay L. 2017. "Are Catastrophic Disasters Striking More Often?" 2017. https:// theconversation.com/are-catastrophic-disasters-striking-more-often-83599.

Zhang, D. D., P. Brecke, H. F. Lee, Y. Q. He, and J. Zhang. 2007. "Global Climate Change, War, and Population Decline in Recent Human History." *Proceedings of the National Academy of Sciences* 104, no. 49 (2007): 19214–19. https://doi.org/10.1073 /pnas.0703073104.

Zhang, Zuoyi. 2019. "HTR-PM: Making Dreams Come True—Nuclear Engineering International." 2019. https://www.neimagazine.com/features/featurehtr-pm -making-dreams-come-true-7009889/.

Zhu, Zaichun, Shilong Piao, Ranga B. Myneni, Mengtian Huang, Zhenzhong Zeng, Josep G. Canadell, Philippe Ciais, et al. 2016. "Greening of the Earth and Its Drivers." *Nature Climate Change* 6, no. 8 (2016): 791–95. https://doi.org/10.1038 /nclimate3004.

INDEX

ABOUT THE AUTHOR

Bjorn Lomborg is the best-selling author of *The Skeptical Environmentalist* and *Cool It,* along with numerous academic books and publications. He is a visiting professor at Copenhagen Business School and a visiting fellow at the Hoover Institution at Stanford. As the president of the Copenhagen Consensus think tank, he has worked with more than three hundred of the world's leading economists and seven Nobel laureates, and helped change billions of dollars in donor spending. His work appears in the *New York Times,* the *Wall Street Journal,* the *Washington Post,* the *Economist,* the *Atlantic,* the *Financial Times,* and many other publications. His monthly column appears in dozens of the leading papers around the world, reaching tens of millions of readers. *Time* has called him one of the hundred most influential people in the world, he has repeatedly been named one of *Foreign Policy*'s Top 100 Global Thinkers, and the *Guardian* has identified him as "one of the fifty people who could save the planet." He lives in Prague.